D0575043

Origins

Origins

Richard E. Leakey
and Roger Lewin

What New Discoveries Reveal About the
Emergence of our Species and its Possible Future

Macdonald and Jane's · London

First published in Great Britain in 1977 by
Macdonald and Jane's Publishers Limited
Paulton House, 8 Shepherdess Walk, London, N1

This book was designed and produced by
The Rainbird Publishing Group Limited
36 Park Street, London, W1Y 4DE

Designer: Ruth Prentice

ISBN: 0 354 04162 2

Text photoset by Jolly and Barber Ltd., Rugby,
Warwickshire, England
Printed and bound by Dai Nippon Printing Co.,
Tokyo, Japan

Contents

1 Humanity in Perspective

Close to three million years ago on a campsite near the east shore of Kenya's spectacular Lake Turkana, formerly Lake Rudolf, a primitive human picked up a water-smoothed stone, and with a few skilful strikes transformed it into an implement. What was once an accident of nature was now a piece of deliberate technology, to be used to fashion a stick for digging up roots, or to slice the flesh off a dead animal. Soon discarded by its maker, the stone tool still exists, an unbreakable link with our ancestors; together with many others, that tool is preserved in the National Museums of Kenya in Nairobi. It is a heart-quickening thought that we share the same genetic heritage with the hands that shaped the tool that we can now hold in our own hands, and with the mind that decided to make the tool that our minds can now contemplate.

There is an inescapable and persistent element of excitement in the search for the origins of humanity. It affects everyone, professionals and non-professionals alike, because there appears to be a universal curiosity about our past, about how a thinking, feeling, cultural being emerged from a primitive ape-like stock. What evolutionary circumstances molded that ancient ape into a tall, upright, highly intelligent creature who, through technology and determination, has come to dominate the world? This is the question we ask about ourselves. And it is not mere idle curiosity because, without doubt, the key to our future lies in a true understanding of what sort of animal we are.

Ever since the first signs of self-awareness flickered in the minds of our distant ancestors the human (or pre-human) mind has pondered on its relationship with the world outside. We can guess that early humans, say a million or so years ago, were conscious of themselves as an integral part of the environment in which they lived: they were hunters and gatherers and they survived only if they respected the world in which they lived. And yet they may have already

Previous page: Our small planet revolves round the sun, which is only one of the ten thousand million stars in our galaxy. This in turn, is one of the millions of galaxies that make up the universe which, according to the 'big bang' theory, probably began with a gigantic explosion about thirteen billion years ago. The enormity of this scale puts our existence on earth in perspective. The illustration is of the Great Galaxy in Andromeda, some seven hundred and fifty thousand light years from us.

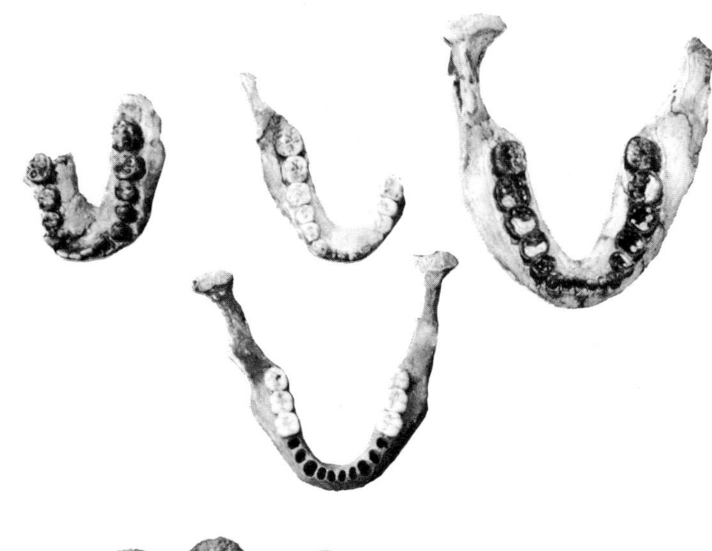

The Leakey family have played a most important part in pushing back the time at which the hominid line first evolved in Africa. Top left, Dr Louis S. B. Leakey, then twenty years old, on a dinosaur dig in 1924. Olduvai Gorge, Tanzania, has yielded a large number of fossil bones. This site, left, is where the remains of Homo habilis *were found. Workmen are screening loose earth from cuts in the hillside that are protected by a tarpaulin. The figures in the background include Drs Louis and Mary Leakey. Above: an aerial view of Olduvai Gorge, on the edge of Serengeti, where so many early finds were made. Above right, the jaws of early man contrasted with modern man and near-man. The photograph shows the lower jaws of a young* Homo habilis *and a possible* Australopithecus africanus *female of about twenty years old found at Olduvai Gorge (left and center) alongside the massive mandible of* Australopithecus boisei, *and (below) present-day* Homo sapiens sapiens. *The* Australopithecus *jaw, center right, found by Kamoya Kimeu at Lake Natron, some fifty miles from Olduvai, has worn premolars, which suggest a rougher, grittier diet. Compared with its other teeth its incisors are small. The fossil jaw of an early* Homo *was found by Dr Mary Leakey at Laetolil in Tanzania. The lateral view, center right, shows permanent teeth below the milk teeth. Right, Richard Leakey working in the field near Lake Turkana in the fall of 1976.*

begun the age-old human practice of attempting to secure more favorable treatment for themselves by appealing in diverse ways to the greater natural forces that rule the world. Time marched on and eventually modern humans – *Homo sapiens sapiens*[1] – emerged, creatures who, to an extraterrestrial observer, must seem to be more than a little perverse. Unlike no other animal, we wage war on each other. We knowingly exploit limited resources in our environment and seem to expect that our profligacy can go on forever. And we choose to ignore deep chasms of injustice, consciously inflicted both between nations and within nations. In a sense it is humans who now rule the world: our extraordinary creative intelligence gives us the potential to do more or less anything we want. But, an extraterrestrial observer may wonder, isn't the ruler just a little bit crazy?

If we are not crazy, and we will assume we are not, why is it that humanity seems determined to spiral ever faster towards self-made destruction? Perhaps the human species is just a ghastly biological blunder, having evolved beyond a point at which it can thrive in harmony with itself and the world around it. That must be a possibility. In recent years scientists, playwrights, and others have attempted to explain why mankind finds itself faced with the prospect of self-annihilation. The idea was proposed that man[2] is unswervingly aggressive, an idea that was given scientific credence by proponents such as Professor Raymond Dart and Dr Konrad Lorenz, and successfully popularized by Robert Ardrey.

The core of the aggression argument says that because we share a common heritage with the animal kingdom we must possess and express an aggressive instinct. And the notion is elaborated with the suggestion that at some point in our evolutionary history we gave up being vegetarian ape-like creatures and became killers, with a taste not only for prey animals but also for each other. It makes a good gripping story. More important, it absolves society from attempting to rectify the evil in the world. But it is fiction – dangerous fiction.

[1] We are using the scientific name *Homo sapiens sapiens* ('modern' man) in order to distinguish him from his early ancestor *Homo sapiens*. All the races living today are members of the subspecies *Homo sapiens sapiens*.

[2] As a convenience of style we use the word man and the term mankind to refer to human beings without distinction to gender.

Unquestionably we are part of the animal kingdom. And, yes, at some point in our evolution we departed from the common dietary habits of the large primates and took to including a significant amount of meat in our menu. But a serious biological interpretation of these facts does *not* lead to the conclusion that, because once the whole of the human race indulged in hunting as part of its way of life, killing is in our genes. Indeed, we argue that the opposite is true, that humans could not have evolved in the remarkable way in which we undoubtedly have unless our ancestors were strongly cooperative creatures. The key to the

The study of primates assists us to unravel much that is important in our biological history Those shown are the slow loris and the gorilla.

The tribal economy of contemporary hunter-gatherers – such as the !Kung of the Kalahari (below right) may furnish many valuable clues to the way of life of our early ancestors.

transformation of a social ape-like creature into a cultural animal living in a highly structured and organized society is sharing: the sharing of jobs and the sharing of food. Meat eating was important in propelling our ancestors along the road to humanity, but only as part of a package of socially-oriented changes involving the gathering of plant foods and sharing the spoils.

This being so, why then is recent human history characterized by conflict rather than compassion? We suggest that the answer to this question lies in the change in way of life from hunting and gathering to farming, a change which began about ten thousand years ago and which involved a dramatic alteration in the relationship people had both with the world around them and between each other. The hunter-gatherer is a part of the natural order; a farmer necessarily distorts that order. But more important, sedentary farming communities have the opportunity to accumulate possessions, and having done so they must protect them. This is the key to human conflict, and it is greatly exaggerated in the highly materialistic world in which we now live.

If we are honest we have to admit that we will never

There are many important sites in Africa other than those at Olduvai and Lake Turkana. In South Africa are Makapansgat (left) and Sterkfontein, where Astralopithecus africanus *was found, a skull of which is at the foot of the page. The Omo Valley site in Ethiopia (far left) has produced* A. boisei, *and further north in Ethiopia is Hadar, where Dr Don Johanson has found fossils of* Homo *and* Australopithecus. *Right and left halves of the former's lower jaw are illustrated below.*

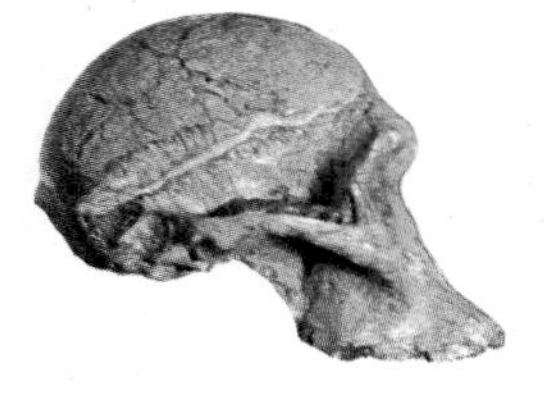

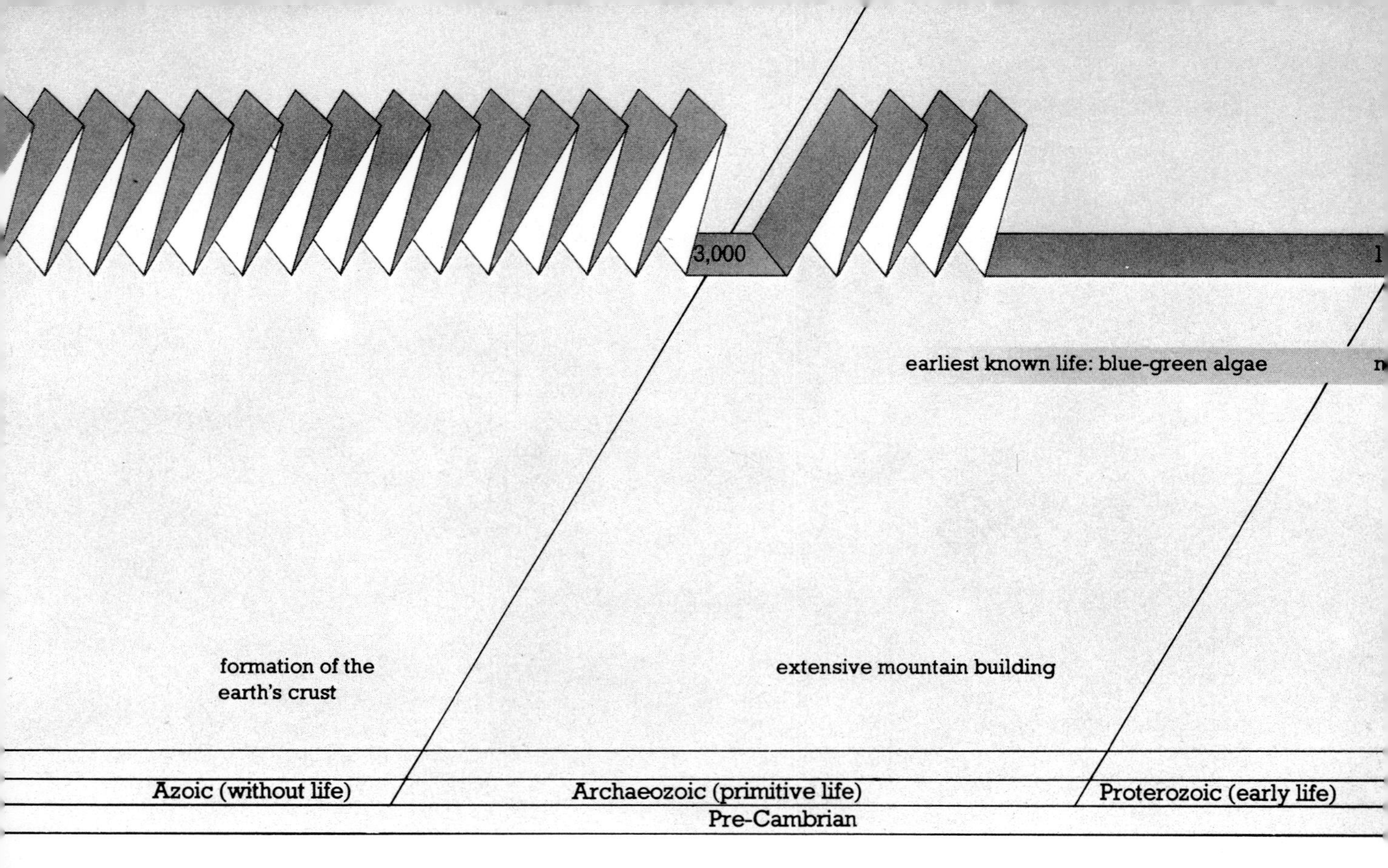

The chart on this and the next opening summarizes the geological and paleontological history of our earth.

fully know what happened to our ancestors in their journey towards modern humanity: the evidence is simply too sparse. For instance, only a small proportion of the campsites occupied two million years ago were preserved by favorable environmental conditions. Of those that did survive only a tiny fraction will ever be discovered, the rest remaining a buried, mute record of the past. And in those occupation sites that are discovered all we can hope to find is a few stone implements and some scraps of fossilized bone, thin testimony to what was once the scene of complex social and cultural activity. Despite limited information about early human culture we are now witnessing a revolution in the study of prehistory. What was once the relatively isolated pursuit of stone-tool and fossil hunters is now the focus of an integrated campaign by scientists of many disciplines.

Any scientific meeting on our origins nowadays might be attended by archeologists (who search for stone tools), paleoanthropologists (who look for early human fossils), geologists (who study the environments of ancient living sites), taphonomists (who investigate the way bones may become buried and subsequently fossilized), anthropologists (who learn the ways of contemporary 'simple' societies), animal behaviorists (who study the habits of monkeys and apes), and psychologists (who may be interested in the development of the human brain). It may seem a motley collection, but by weaving together the threads of their knowledge and expertise a more complete picture than ever before can be created about human origins. Each discipline not only contributes facts which can be incorporated into our understanding of human evolution, but because of the different way in which the varying approaches interact, new questions can be asked which previously were inconceivable.

By good fortune this new approach to human prehistory comes at a time when prehuman fossils are being discovered at an unprecedented rate. During the 1970s, sites in Ethiopia, Kenya, and Tanzania have yielded so many important fossils that it will take years before they will have been through the thorough analysis that is now accorded such specimens. But it is not simply the *number* of fossil finds that is significant and exciting, it is the *nature* of the fossils themselves that is causing so much of a stir among prehistorians. Our

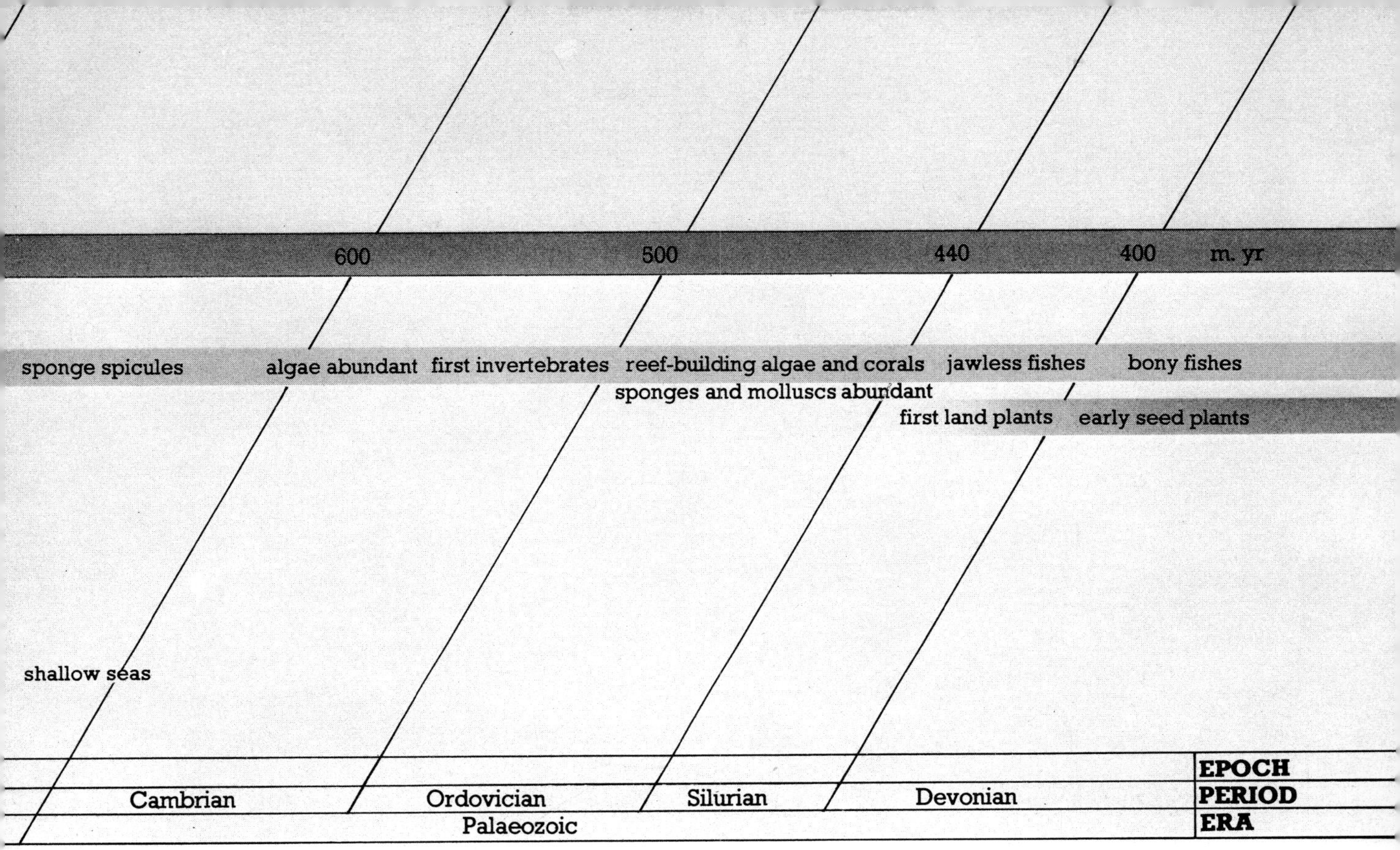

600 500 440 400 m. yr

sponge spicules | algae abundant | first invertebrates | reef-building algae and corals | jawless fishes | bony fishes

sponges and molluscs abundant

first land plants | early seed plants

shallow seas

				EPOCH
Cambrian	Ordovician	Silurian	Devonian	PERIOD
Palaeozoic				ERA

view of human evolution has been transformed within the past few years. It is now apparent that the ancestral line that led to modern humans stretches back five, perhaps six, million years. And it is clear that for a large part of that time our ancestors shared their environment with two types of creatures with whom they were closely related but who eventually became extinct. These evolutionary cousins are called the australopithecines, one of whom was slightly built while the other was much more robust.

The two forms of australopithecine and the *Homo* ancestor shared at least two things: first, they shared a common ancestor, a small ape-like creature called *Ramapithecus* who first appeared at least twelve million years ago and lived in Europe, Asia, and Africa; and second, they all stood and walked upright. At present we cannot be certain what evolutionary pressures allowed *Ramapithecus* to diversify into its australopithecine and *Homo* descendants, a process which, incidentally, appears to have happened only in Africa and not in any of the other continents in which it lived. But, because of the new approach to the study of mankind's past and the flood of fossil discoveries, we can begin to guess at the subtle differences in behavior which separated these three evolutionary cousins, known collectively as hominids. At first their ways of making a living would not have been dramatically different. But gradually the growing social complexity of the *Homo* stock would have driven a bigger and bigger evolutionary wedge between it and the australopithecines. The adoption of the combination of meat eating and plant-food gathering was a vital part of that upgrading in social organization, and it was a way of life that dominated human existence until a mere 10,000 years ago when people began to exploit the potential of farming.

The story since then is well known: the Agricultural Revolution was followed by the Industrial Revolution (last century), and that led to today's technological revolution. And with these dramatic changes in the lifestyle of human beings has come an explosion in world population: from something less than ten million on the eve of the Agricultural Revolution, it currently stands in excess of four thousand million, two thirds of whom are starving!

If geographical distribution is a signal of success, then man is successful: from our evolutionary cradle in Africa we now occupy virtually every corner of the globe. Wherever life is possible, there is man. Because of the unusual adaptability that was a crucial

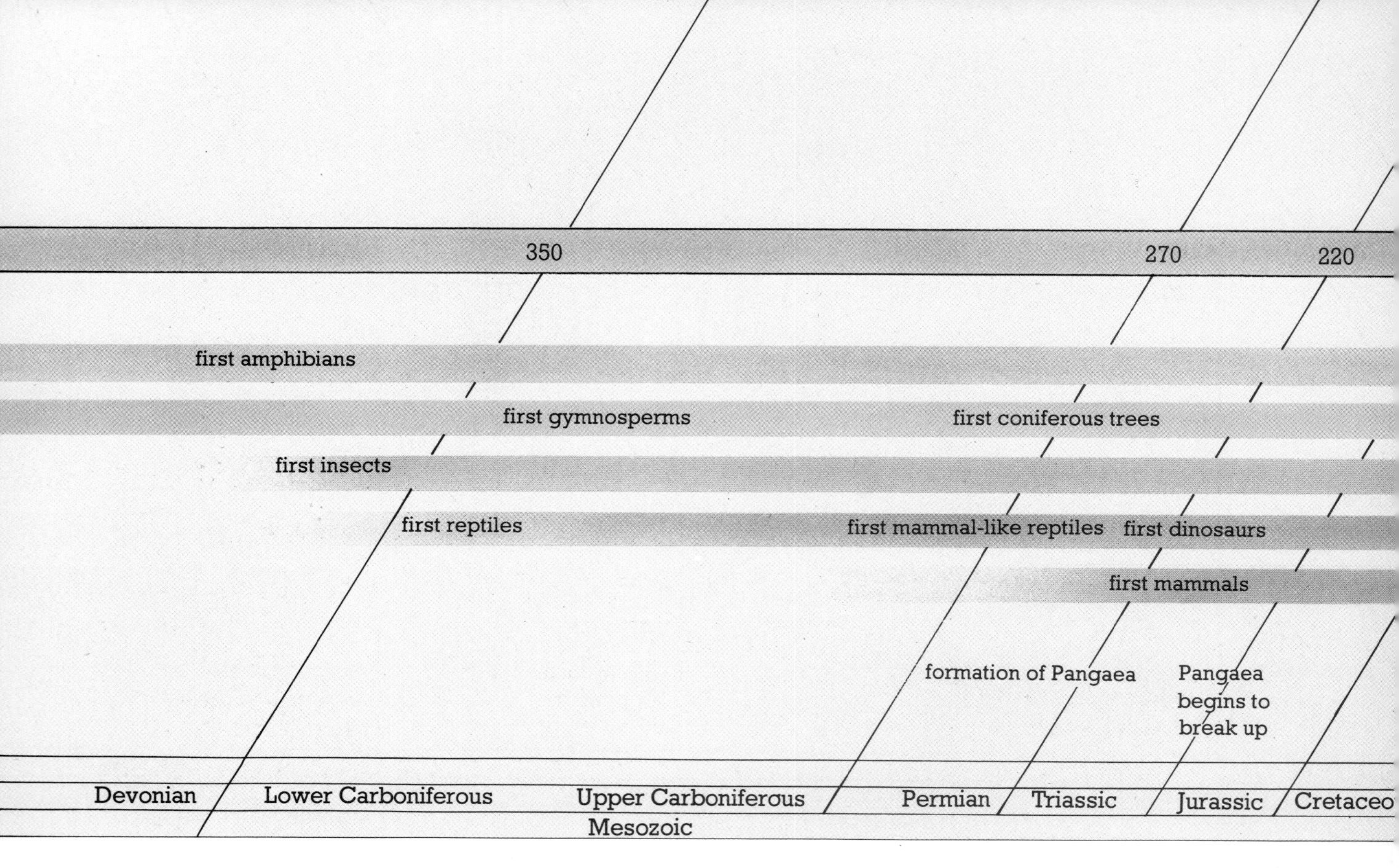

element in our evolution, no other single species inhabits such a variety of environments.

We should, however, put man's unusual ascendancy into perspective. If we were to document the history of the earth, day by day, year by year, since its birth as part of the solar system some four and a half thousand million years ago in a single volume of exactly a thousand pages, each of those pages would cover four and a half million years. Almost the first quarter of the book, about 220 pages, would describe how conditions propitious for the emergence of life slowly came about after the gases had condensed to form our hot seething planet. At this point, blobs of jelly, unmistakably living, yet very primitive, would be seen in the swirling tide pools of the warm oceans. But life in the sea in a form with which we are familiar would have to wait until we plowed our way through three quarters of the text – the Age of Fishes was 500 million years ago. And the first land creatures, descendants of fishes that deserted their aquatic habitat, turn up 30 pages later, at about 350 million years ago. One of the most exotic periods, and certainly the most awe-inspiring, of the earth's history, the Age of Dinosaurs, would consist of about 30 pages, describing the period between 225 million and 70 million years before the present, when, with unusual abruptness, they disappeared, to be replaced by the Age of Mammals. It was at this point, 70 million years ago, that the first primates evolved, small rat-like creatures that abandoned ground living and took to life in the trees: it was from such simple beginnings that monkeys, apes, and humans evolved – we share a primate heritage with the monkeys and apes, and with smaller creatures too, such as the mouse lemur and the diminutive potto.

The most distant of man's identifiable ancestors (the first hominid) puts in an appearance about three pages from the end of the book, at around twelve million years: this was *Ramapithecus*. The *Homo* lineage comes at the bottom of the penultimate page, and the first stone tools would be described half way down the last page. And, testing our powers of literary compression to an extreme degree, the whole rise of modern humans would have to be crammed into the last line of the book, with the esthetics and symbolism of the stone-age cave paintings, the advent of agriculture, the intellectual excitement of the Renaissance, the turbulence of the Industrial Revolution, the polarization of the Superpowers, the birth of space travel, and everything else that constitutes our recent history, somehow telescoped into the final word!

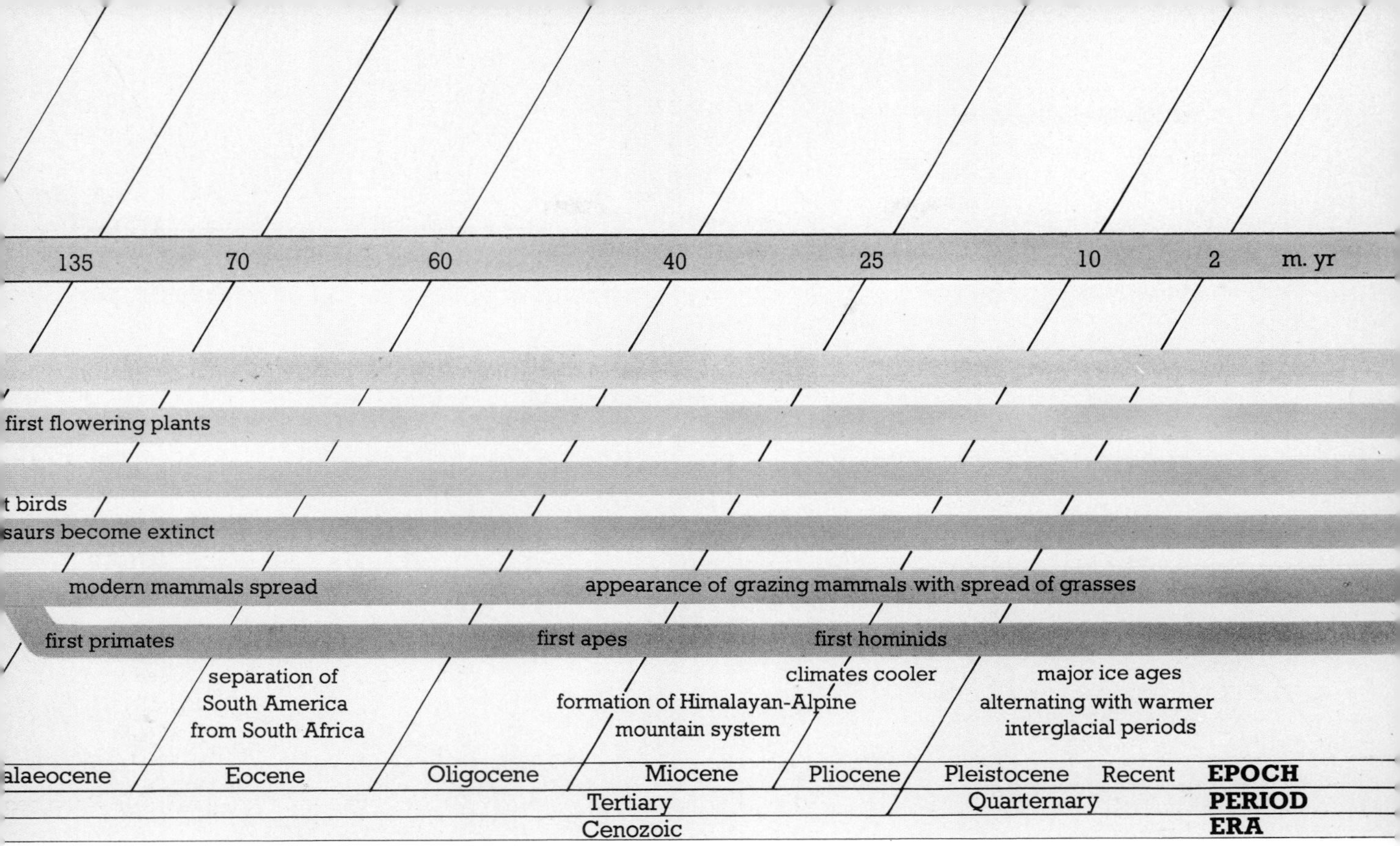

Humans may be masters of the earth for the moment, but we should reflect that just as our planet has had a long history, of which modern humans have occupied a tiny fraction of one per cent, so too does it have a long future. In another two hundred million years will *Homo sapiens sapiens* still have a leading role?

The answer is almost certainly, No – for two reasons. First, if we continue to display the arrogance and profligacy that mark the behavior of so-called civilized people, we will soon have taxed the environment beyond the limits of our own adaptability, if not that of the earth itself. And second, as our condensed history of the earth shows, long-term stability for a single highly-complex species, no matter how adaptable, is biologically out of the question. In many ways our unique culture transcends our biology, and for that reason our future becomes even more uncertain than it might otherwise have been. And the power of culture is such that, rather than question the possible status of our species two hundred million years hence, we had better be concerned about developments in a tiny fraction of that time. For the human species today two hundred years, let alone two hundred million years, is a very uncertain prospect. Now, more than ever before, we depend on the versatility of our culture to shape our daily lives. Similarly, it is upon the flexibility and strength of our culture that we shall have to rely for the future security of our kind. Through culture we have the power to create a future either of justice and compassion, or of suffering and misery. Culture has endowed us with that much choice.

By searching our long-buried past for an understanding of what we are, we may discover some insight into our future. There is much more than bones and stones buried in those fossil-bearing sediments: there are vital clues to human biology. Through an exploration of the forces that nurtured the birth of the hunting and gathering way of life perhaps three million years ago, and through studying the question of why such a long-established mode of existence was superseded, beginning some 10,000 years ago, by a sedentary agricultural society, we can hope for some insight into modern society, and with it some guide to our future. That is the aim of this book.

Overleaf: A map of the chief sites in the world on which our knowledge of man's fossil past is based. The proliferation of sites in Europe is almost certainly a reflection of the fact that, at least in earlier times, so much work was done there.

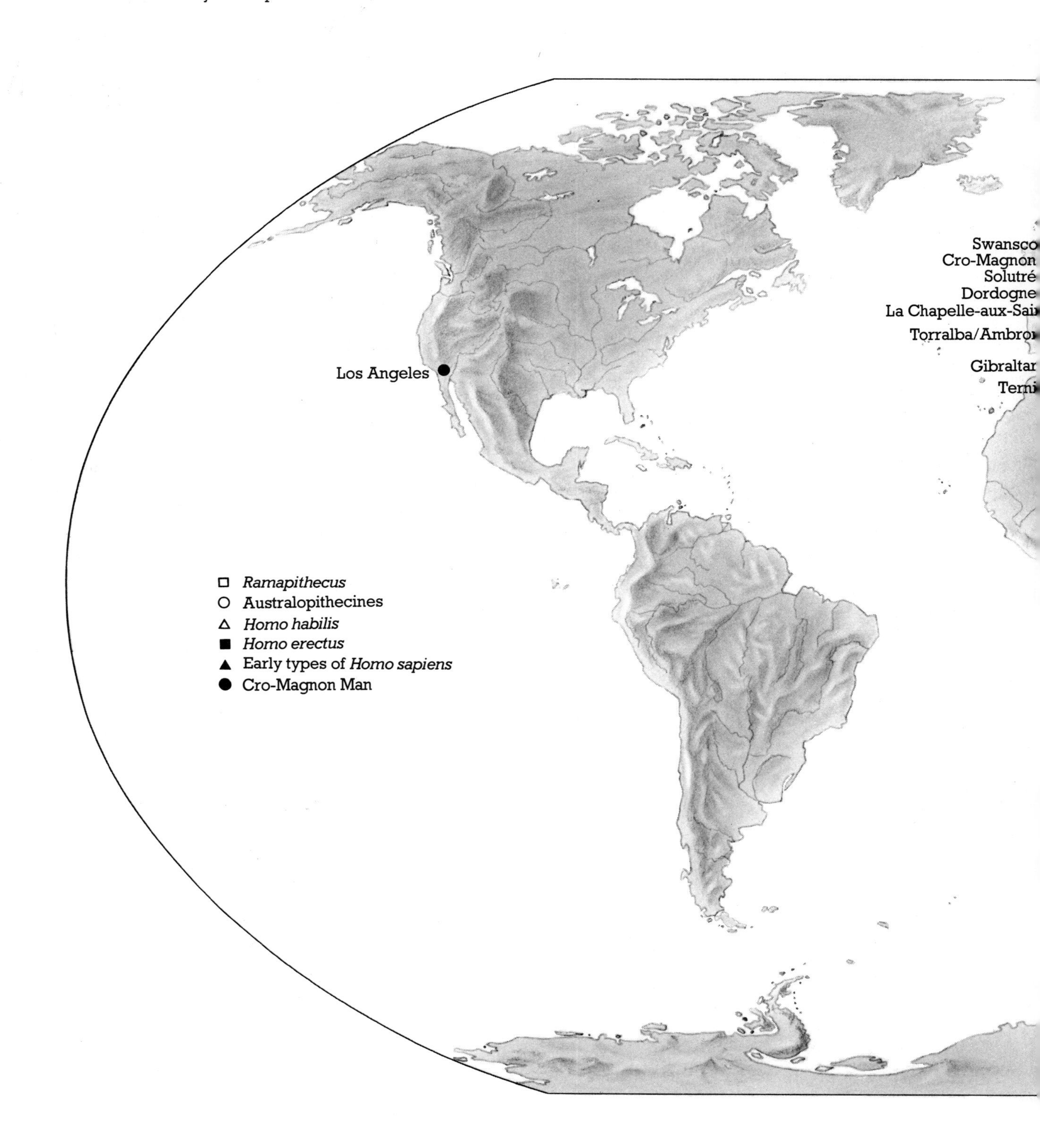
Swansco
Cro-Magnon
Solutré
Dordogne
La Chapelle-aux-Sai
Torralba/Ambro
Gibraltar
Terni
Los Angeles
Ramapithecus
Australopithecines
Homo habilis
Homo erectus
Early types of Homo sapiens
Cro-Magnon Man

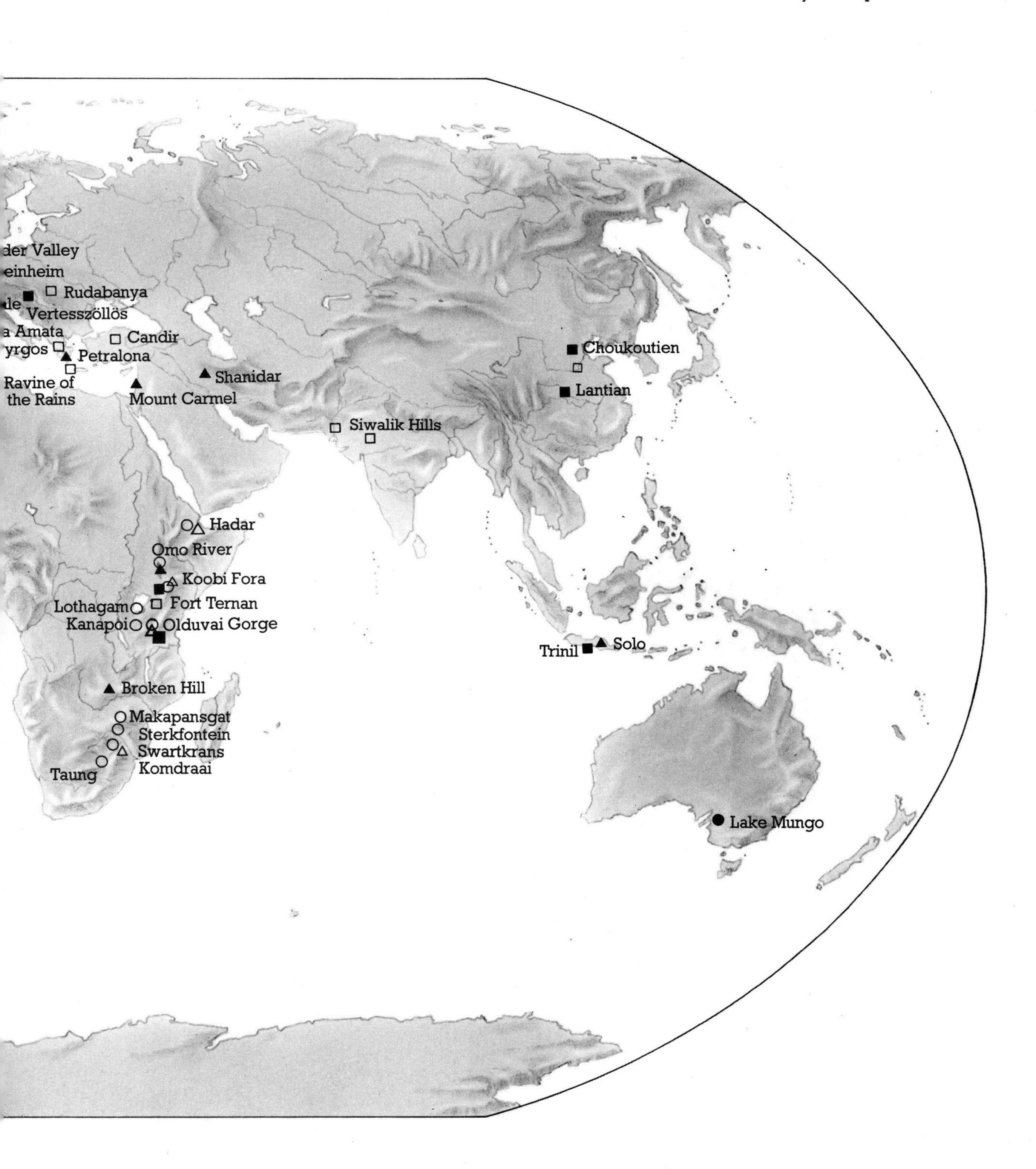
der Valley
einheim
le
Rudabanya
Vertesszöllös
a Amata
yrgos
Candir
Petralona
Ravine of
the Rains
Mount Carmel
Shanidar
Choukoutien
Lantian
Siwalik Hills
Hadar
Omo River
Koobi Fora
Fort Ternan
Lothagam
Kanapoi
Olduvai Gorge
Broken Hill
Makapansgat
Sterkfontein
Swartkrans
Komdraai
Taung
Trinil
Solo
Lake Mungo

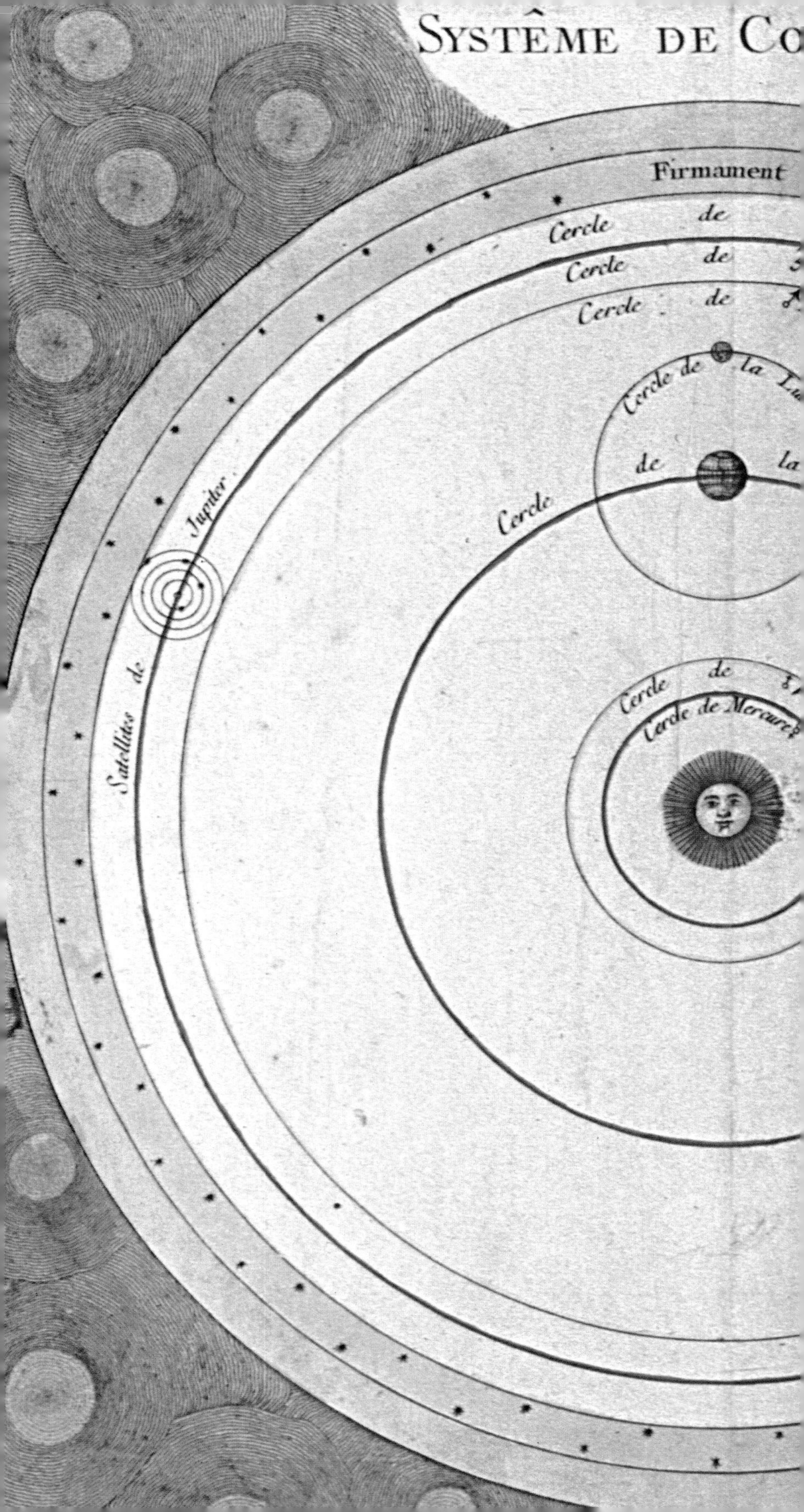

SYSTÊME DE CO

Firmament

Cercle de

Cercle de

Cercle de

Cercle de la Lu

Cercle de la

Satellites de Jupiter

Cercle de

Cercle de Mercure

2 The Greatest Revolution

On hearing, one June afternoon in 1860, the suggestion that mankind was descended from the apes, the wife of the Bishop of Worcester is said to have exclaimed, 'My dear, descended from the apes! Let us hope it is not true, but if it is, let us pray that it will not become generally known.' As it turns out, she need not have been quite so worried: we are *not* descended from the apes, though we do share a common ancestor with them. Even though the distinction may have been too subtle to offer her much comfort, it is nevertheless important.

The question, indeed, was no less than the second of two major intellectual revolutions through which humanity has had to come to terms with its place in the natural world. The first occurred more than four hundred years ago, when the Polish mathematician Nicolaus Copernicus shattered the notion that the earth is the center of the universe. The second began to erupt when Charles Darwin showed that mankind was part of nature rather than apart from it.

Goethe once declared that 'of all the discoveries and opinions proclaimed, surely nothing has made such a deep impression on the human mind as the science of Copernicus.' Although one would be hard put to choose between the two discoveries, it would have been interesting to hear Goethe's opinion had he lived to witness the impact of the Darwinian revolution. Certainly the science based on Darwin's notion of a steady progression of more and more complex organisms as a result of natural selection has a legitimate claim to being the greatest intellectual and philosophical revolution in human history.

For almost two millennia the Judeo-Christian story of the Creation was taken for granted throughout the Western world. With no good reason to doubt it, the teaching of the increasingly powerful Christian churches that God created man in his own image was a comfortable one. There was a certain curiosity, though, about just when this miraculous event had occurred. James Ussher (1581–1656), Archbishop of Armagh, came up with an answer in 1650, when he announced, as a result of his calculations based on the numerology of the Old Testament, that the Creation had taken place in 4004 B.C. Ussher's calculations were later given even greater precision by Dr John Lightfoot, Master of St Catherine's College, Cambridge, England, who declared the precise day to be 23 October, and the time exactly nine o'clock in the morning. Along with an impressive though dubious

Pages 18 and 19: Regarded by many as the founder of modern astronomy, Nicolaus Copernicus, Polish astronomer, mathematician and man of God, published his theory of the heliocentric solar system in 1543, putting our planet and therefore ourselves in true cosmological perspective. This illustration of his views is from a print published in Paris in 1761.

Opposite: 'God Creating the Universe' by a French mid-13th century painter is typical of the standard view held by so many for so long.

Archbishop of Armagh, James Ussher, noted for his firm statement of biblical chronology that included 4004 BC as the year of the Creation. Engraving by George Vertue after the portrait by Sir Peter Lely.

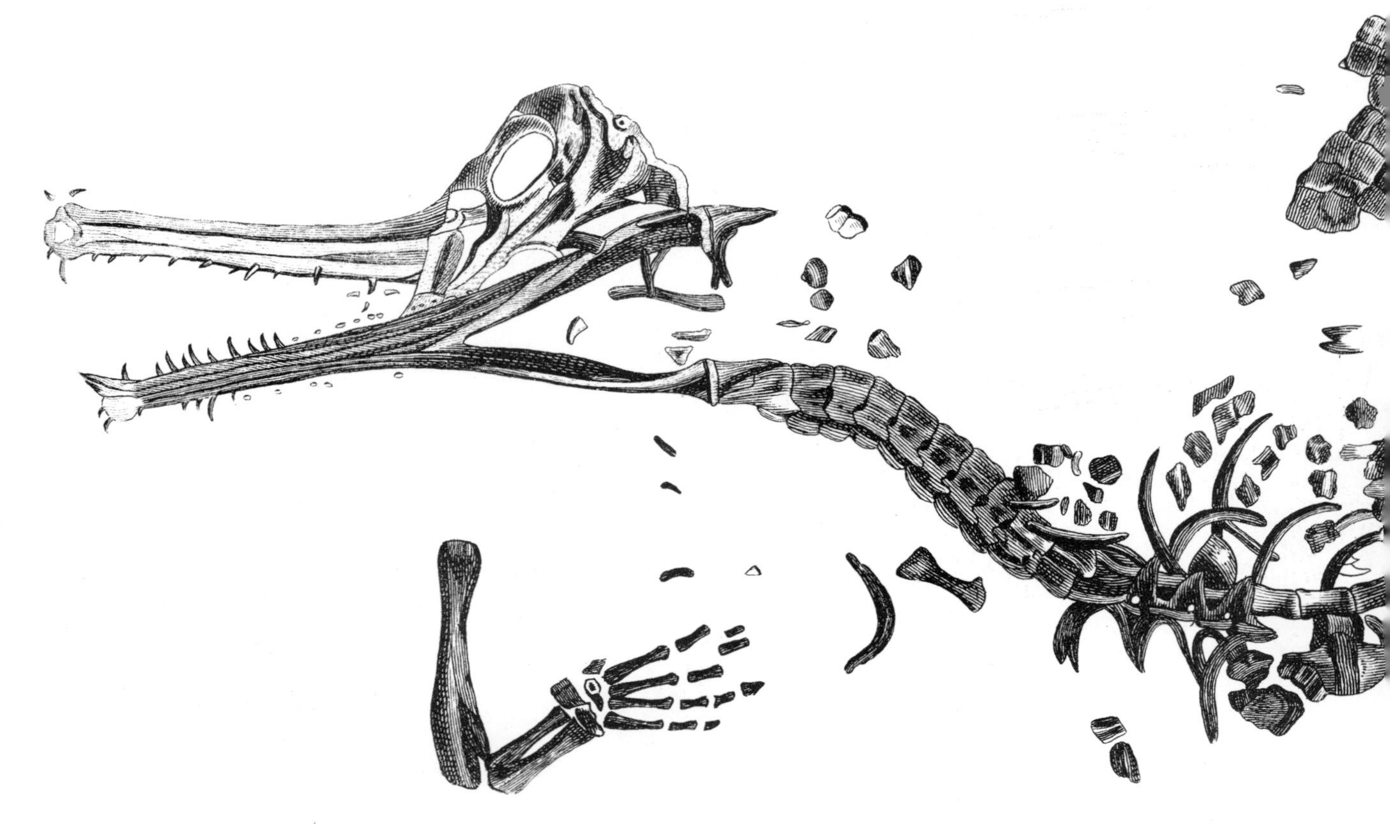

chronology, Lightfoot's dating showed a tender concern for himself and his colleagues, as it coincided with the beginning of the academic year!

The Usher/Lightfoot calculation thus gave the earth a past of a modest six thousand years. Although the extraordinary longevity of many of the Old Testament characters ought to have presented a few problems if it had been seriously contemplated, the six thousand-year period was generally accepted until evidence to the contrary started to turn up. Amateur geologists, many of whom were also clergymen, began to be puzzled by the stratified nature of many rocks, implying a process of formation that was believed not to be operating in contemporary times, but which was somewhat difficult to reconcile with received accounts of the Creation. And, to make matters worse, there were fossils too, and these were clearly the remains of animals no longer in existence.

During the eighteenth century, as evidence of this kind became overwhelming, an explanation was sought that would still be consonant with the teachings of the Bible. The result was the Diluvial Theory, which accounted for fossils by stating that they were the remains of animals which had perished in the Noachian Flood.

Soon, however, the Diluvial Theory came under pressure, as it became clear that a single event like the Flood could not explain the apparent progression of fossils in different layers of rocks, those of each layer more primitive than in the one above it. The final blow came with the discovery of 'pre-Flood' fossils that were clearly related to animals living after the Deluge.

Meanwhile, the view of human origins remained essentially the same as had been laid down in the Old Testament: that humanity had appeared all at once, in the fully modern shape of Adam and Eve. As the steady trickle of ancient stone and flint tools, at first discounted as works of nature or as thunderbolts, went on being unearthed, questions began to stir in some enlightened minds. The first recorded suggestion of a great antiquity for man came in 1797, when John Frere in a paper delivered to the Royal Society in London described flint tools discovered under twelve feet of earth at Hoxne, near Diss in Suffolk. Frere who, incidentally, was Mary Leakey's great great great grandfather, opined that the tools 'were fabricated

Opposite: This fossil crocodile, illustrated in Cuvier's book, The Animal Kingdom *(1830), is obviously related to present-day species and it was such finds that posed a problem to the proponents of the Diluvial Theory.*

Baron Georges Leopold Cuvier, the French comparative anatomist, explained away the progressive sequences of fossils found in strata by proposing a series of catastrophes, the Flood being just one of these.

and used by a people who had not the use of metals . . . . The situation [depth] at which these weapons were found may tempt us to refer them to a very remote period indeed, even beyond that of the present world.' Frere's insight – and courage – apparently went unheeded, until the mid-nineteenth century, when it was adduced to demonstrate Britain's accomplishments in archeology!

Meanwhile, orthodox Christianity was saved from the embarrassing inadequacies of the Diluvial Theory by the French geologist, naturalist, and member of the Académie des Sciences, Baron Georges Cuvier (1769–1832). To explain the progressive sequences of fossils found in rock sediments, Cuvier proposed a series of catastrophes, each of which had totally wiped out animal and plant populations (thus producing the fossils), followed by a period of calm during which God restocked the earth with new (and improved) species. The Noachian Flood was just one of these.

The Catastrophe Theory was a great balm to many troubled minds. Adam Sedgwick, a geologist at Cambridge University and a teacher of Charles Darwin, expounded the theory thus: 'At succeeding periods new tribes of beings were called into existence, not merely as progeny of those that had appeared before them, but as new and living proof of creative interference; and though formed on the same plan, and bearing the same marks of wise contrivance, oftentimes unlike those creatures which preceded them, as if they had been matured in a different portion of the universe and cast upon the earth by the collision of another planet.'

In formulating the Catastrophe Theory, Cuvier routinely took for granted an extreme rapidity of changes in times past as compared with the present, but conceded that perhaps a little more than six thousand years was required. So, following the example of his countryman, Comte Georges de Buffon (1707–1778), he added eighty thousand years on to the age of the earth. According to calculations of members of the Académie, made after Cuvier's death, there had been twenty-seven successive acts of creation, the products of each but the last being obliterated in subsequent catastrophes, thus providing a geological 'clock'. An Englishman, William Smith (1769–1839), raised the number of strata to thirty-two.

It is evident that a major problem throughout this period of debate concerning the origin of fossils was that of attributing a correct age to the earth and the strata from which fossils were retrieved. The first real pioneer in the effort to solve the problem was a Scotsman, James Hutton (1726–1797), who concluded from a study of the available geological evidence that the forces that had shaped the world in the past, built mountains and created continents, were still at work in the modern world. He saw the life of the earth as a continuum, rather than as a past and a present neatly divided by a line that marked the creation of man by God. Hutton, though not the first to suspect that the world was far more ancient than had been generally supposed, was the first writer to set forth a coherent argument for such a view – his *Theory of the Earth*. Published in 1795, when it was met with scorn and ridicule, it proposed a thesis that came to be known as Uniformitarianism. Hutton died just two years later, in 1797 – a year that saw the birth, to well-to-do parents in Scotland, of Charles Lyell. And it was Lyell who revived and established Uniformitarianism as a theory without which the achievement of Charles Darwin would have been impossible.

In 1830 Lyell published the first volume of his monumental *Principles of Geology*, and thus became the father of modern geology. In part his work was one of synthesis, documenting in meticulous detail the evidence leading to the inescapable conclusion that *Homo sapiens* inhabited a planet of great antiquity.

The message of his *Principles* was not received without opposition, however. Adam Sedgwick in England, and Cuvier in France, as convinced Catastrophists, aligned themselves against it. Meanwhile, ironically, some of the most impressive evidence supporting the antiquity of mankind was being unearthed

Above: James Hutton – from Original Portraits and Caricature Etchings *by John Kay – the Scottish geologist who put forward the doctrine of Uniformitarianism.*

Right: The first volume of Principles of Geology *by Sir Charles Lyell, whose portrait is shown here, was a gift from Professor J. S. Henslow to Darwin on the latter's departure in H.M.S.* Beagle. *He received the second volume while in Montevideo and it is probably true to say that without this work Darwin's later achievement would have been impossible.*

Above: The French naturalist Jean Baptiste de Lamarck, shown here in an engraving after the painting by Charles Thévenin, was a interested in evolutionary theory. He proposed that environmental changes caused changes in the structure of both animals and plants.

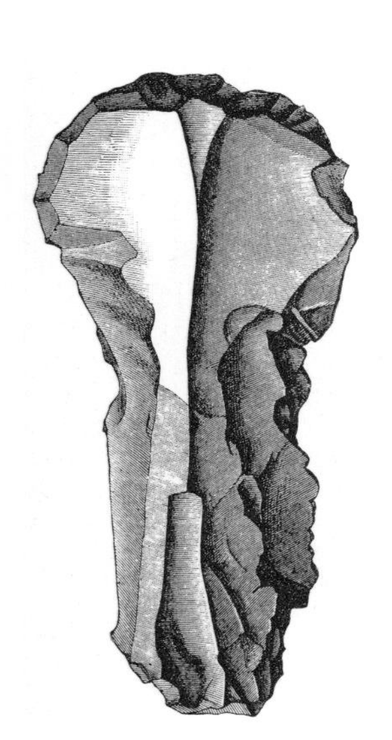

Left: An engraving (slightly reduced) of a flint scraper found at Abbeville in the Diluvium deposits.

at Abbeville, in the northwest of France. There, Jacques Boucher de Crevecoeur de Perthes was excavating stone implements together with the fossilized bones of extinct animals, a contemporaneity which any theory involving the Deluge could not explain.

In the intellectual schism that opened during the eighteenth century between the Creationists and those who believed in some kind of evolution, Erasmus Darwin (1731–1802), Charles's grandfather, was a pioneer and principal spokesman on the side of the evolutionists. Physician, philosopher, poet, and celebrated personality, in his writings between 1784 and 1802 Erasmus Darwin posed two questions: first, whether all living creatures are ultimately descended from a single common ancestor; and second, how species could be transformed. To answer the first question he assembled evidence from embryology, comparative anatomy, systematics, and geographical distribution, assimilating the fossil data on the way, for a single source of all life, 'one living filament,' an evolving web of life that included mankind. This after all was in keeping with the eighteenth-century classification of all animals and plants into families, genera and species by the Swedish botanist Carolus Linnaeus (1707–1778), who classed *Homo sapiens* as a close relative of the Old World monkeys and the apes – although scientists and theologians alike had exerted great efforts to extricate mankind from this unseemly association!

The second question – concerning the forces through which evolution is achieved – was trickier to deal with. It is, however, fair to say that Erasmus Darwin's treatment of the problem contained at least the seeds of almost all the important principles of evolutionary theory. He saw that competition and selection were possible agents of change; that over-population was an important factor in sharpening competition; that plants should not be left out of evolutionary theory; that competition between males for females has important structural implications in their evolution; and that fertility and susceptibility to disease were areas of selection. He did not state definitively, however, that the principal agent of evolution is passive adaptation through natural selection, but seemed to admit the possibility that animals may evolve through active adaptation to their environment, including the inheritance of acquired characteristics.

It remained for Jean Baptiste de Lamarck (1744–1829) to take up Erasmus Darwin's mention of

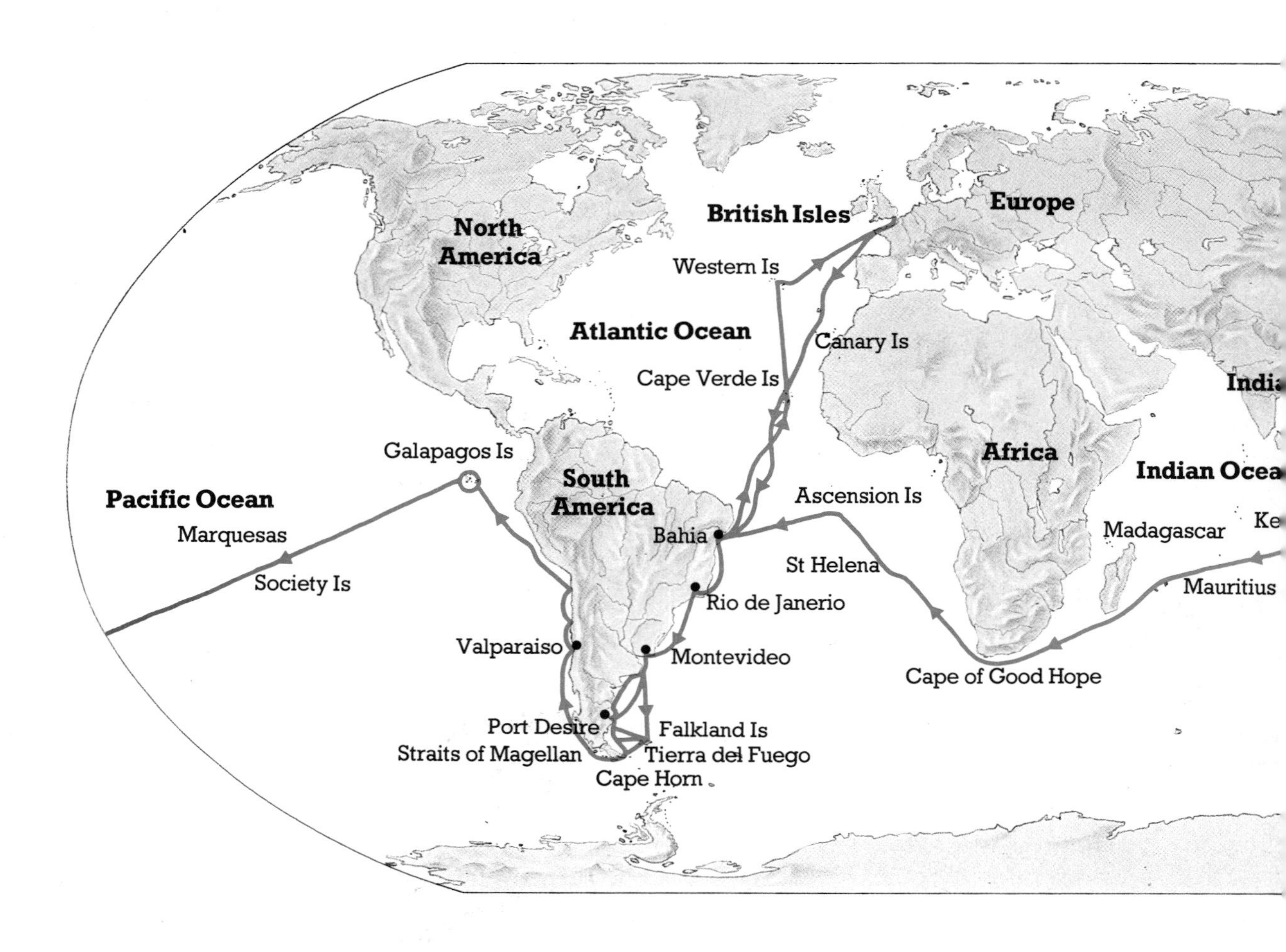

Four of the several species of Galapagos finches engraved to illustrate Darwin's *Journal of Researches*. His observations on the species found in these islands were probably the most important he ever made. It later became clear that the islands had been originally populated by a single ancestral line that had flown from mainland South America. Rapidly filling the available space, competition for territory and food brought about adaptations allowing different populations to live in varying habitats and to feed on different foods. The upper two species are ground-dwellers which feed on seeds, while the lower pair are tree-dwellers and feed on insects. The beak sizes are dependent upon the sizes of the seeds and insects upon which they feed.

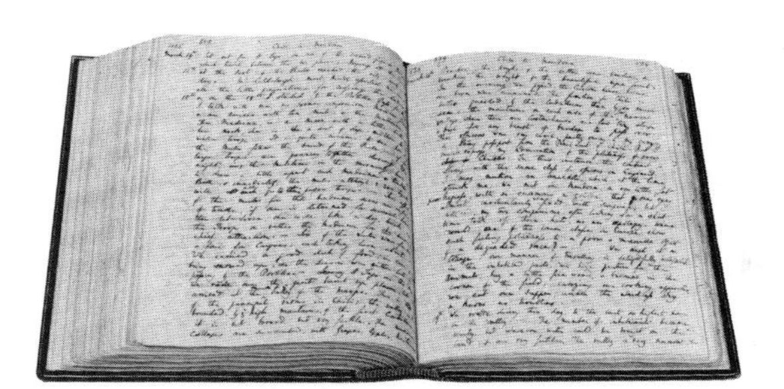

Japan
China
Australia
Sydney
Bay of Islands
Hobart
King George's Sound

Above left, a map showing H.M.S. Beagle*'s ports of call during its five-year voyage around the world and, above, the original diary kept by Darwin, opened at the pages dated 14–19 March, 1835. Right, a water-colour portrait of Charles Darwin painted by George Richmond some four years after the completion of his voyage and, left, one of the many natural history paintings made to illustrate the official* Zoology of the Voyage of H.M.S. Beagle.

inheritance of acquired characteristics and expand it into a fully-fledged theory of evolution. In the process, Lamarck unwittingly exposed the absurdity of supposing that the giraffe's long neck, for example, was simply the result of generations of neck stretching, with the result that Lamarckism, as his theory was dubbed, brought the entire cause of evolution into some disrepute. In 1813, three men – William C. Wells (an expatriate American), James C. Pritchard, and William Lawrence – presented independent rebuttals of Lamarckism to the Royal Society in London. All three of these papers upheld the view of natural selection foreshadowed by Erasmus Darwin, as the engine of evolution. According to Pritchard, 'All acquired conditions of the body end with the life of the individual in whom they are produced.' But for this salutary law the universe would be filled with monstrous shapes.

Inevitably, these propositions were soon the target of sharp attack, particularly from the Church and associated establishments. Nevertheless, the papers of Wells, Pritchard, and Lawrence were read with great interest by contemporary naturalists, and Charles Darwin himself must later have been aware of them. Meanwhile, Charles Naudin, a Frenchman, had been impressed with the structural changes that could be induced in domesticated crops and animals through selective breeding, and reasoned that perhaps a similar process might be operating in nature, but passively.

Yet another of Charles Darwin's intellectual antecedents was Edward Blyth, who as a young man had been much impressed by Lyell's ideas. In 1835 and 1837 he contributed articles to the *Magazine of Natural History*, a magazine with which Darwin was familiar. 'Among animals which produce their food by means of their agility, strength or delicacy of sense,' Blyth wrote, 'the one best organized must always obtain the greatest quantity; and must, therefore, become physically the strongest and be thus enabled, by routing its opponents, to transmit its superior qualities to a greater number of its offspring.'

Such ideas, with the principles of Lyell in the background, came to the very edge of Darwinism. But it remained for the man himself to assemble all the data and to construct an unassailable theory. For Charles Darwin, like Lyell, was an effective synthesizer of existing information; his theory was not entirely new, but he presented it to the world at a time when the intellectual climate was at its most favorable. Moreover, Darwin collected a vast amount of data of his own to buttress the theory against the inevitably skeptical reception.

Charles Darwin was born in 1809 in the pleasant English country town of Shrewsbury. His father, Robert Waring Darwin, was a medical doctor and a devoutly religious man. Not unnaturally Charles was steered towards the medical profession, and he went to Edinburgh University to read medicine. Soon, however, he realized that the profession was not for him, and so – again guided by his father's influence – he went to Cambridge to read divinity. As Charles remarked of his father, 'He was very properly vehement against my turning into an idle sporting man.'

At Cambridge, as at Edinburgh, Charles was no academic prodigy. Along with his interest in such 'idle sports' as shooting, he had a passion for natural history. At both universities his friends were mainly botanists and geologists – among them Adam Sedgwick and the botanist J. S. Henslow, a professor at Cambridge.

It was Henslow who later was responsible for obtaining for Darwin a post on H.M.S. *Beagle*, the ship aboard which Darwin was to amass the evidence that would form the bedrock of his theory. When Charles's father heard of the expedition, he at first forbade his son to join it, but relented following the intervention of Charles's uncle, Josiah Wedgwood. So, two days after Christmas in 1831 Charles Darwin, equipped with a Cambridge degree in theology, Euclid and the classics, but with no qualifications in science, set out as naturalist on the *Beagle* on a five-year voyage around the world. Darwin was following the tradition of the gifted amateur, learning his skills from professionals around him. Thus, equipped with what his uncle Josiah called 'an enlarged curiosity,' this quiet, reticent man set out on the voyage that was to launch possibly the greatest revolution in humanity's concept of itself.

The journey took Darwin first to South America, where he visited many places along the coast, then via the Galapagos Islands to New Zealand and Australia, and home again, calling in at South African ports on the way. The *Beagle* arrived back at Falmouth, England on 2 October 1836. Throughout the voyage, Darwin went through periods of illness, including two weeks of seasickness at the beginning. But wherever the ship halted, Darwin collected samples in profusion: rocks and fossils, birds, insects, and bigger animals too, putting his skills as a taxidermist to good effect.

For Darwin, probably the most significant part of the entire voyage, and arguably even of his whole life, was the four weeks he spent exploring the Galapagos Islands, a lonely Pacific archipelago several hundred miles due west of Ecuador. There he noticed that each island appeared to have its own type of finch. More than that, different ecological niches on a single island were often inhabited by different finches. And yet they clearly all came from a common stock. Specimens of each were added to his collection, which by the time of his return to England was the most comprehensive ever collected by one man.

What he had seen on his long voyage had convinced Darwin that species were not immutable, that they were capable of transformation. The question that remained was – how? Meanwhile, he had a prodigious amount of work to do, and set about it with enthusiasm. Within six months of his return he had sorted out his specimens with the help of Sir Richard Owen, who was known as the British Cuvier, and had had them described by appropriate experts for the official *Zoology of the Voyage of the Beagle*, which was under Darwin's general editorship. He also wrote his own fascinating general account of the voyage, the classic *Journal of Researches*. There followed three more books, *The Structure and Distribution of Coral Reefs* (1842), *Volcanic Islands* (1844), and *Geological Observations of South America* (1846). His published output on these topics is in marked contrast to his reluctance to broach on paper the subject of evolution. Some of that reluctance may be traced to a blunder he made while he was secretary of the Geological Society (1838–1841). A set of mysterious rock formations at Glenroy, Scotland, which Darwin identified as ancient marine beaches cut off from the sea by subsidence, turned out to have been carved by glaciers. The mistake had wounded Darwin's pride. He was not going to be wrong in print again.

Within fifteen months of beginning to set down notes for *The Transmutation of Species* in 1837, he was more than ever convinced that species did change, and he now believed that selection was the key. He had seen how selective breeding with crops and stock brought about basic changes in the organisms; but, as he put it, 'How selection could be applied to organisms living in a state of nature remained for some time a mystery to me.' A flash of insight that was to illuminate the whole problem for him came on 3 October 1838, while he was reading 'for amusement' the book on population by

Naturalist and explorer Alfred Russel Wallace, who, independently of Darwin, formulated the theory of natural selection while in the Malay Archipelago.

Thomas Malthus (1766–1834), asserting that populations tend to increase geometrically unless constrained. Here, it came to him, was the answer: changes that favored an individual would allow it to prosper as compared with others not possessing these new properties; populations of animals with such advantageous mutations thrived, while those with less advantageous traits declined.

It was not until 1842 that Darwin allowed himself 'the satisfaction of writing a very brief abstract' (it was thirty-five pages long) of his theory. A more extended version, amounting to some 230 pages, followed two years later. Then, in 1846, Darwin turned away from the theory of evolution and devoted himself to a study of barnacles, the product of which was four monographs.

Midway through 1856, urged on by his friends Charles Lyell and Joseph Hooker, Darwin began his magnum opus, to be entitled simply *Natural Selection*. Two years later Darwin had completed ten chapters and was well into the eleventh, on the subject of pigeons. But on 8 June 1858 he received a letter that shattered his plans. It was from the naturalist-explorer Alfred Russel Wallace, who knew of Darwin's interest in evolution. In February of that same year, during an expedition to the island of Ternate in the Moluccas, between New Guinea and Borneo, Wallace had been in bed with a fever. As he lay there, tossing and restless, he had been thinking about the problem of *how* species might be transmuted. He too had read Malthus, and he now experienced a sudden flash of insight sparked by that same theory. This was the news conveyed by the short note that arrived on Darwin's desk some four months later. Accompanying it was a twelve-page summary of Wallace's ideas on evolution. They paralleled Darwin's exactly. The fears that Lyell and Hooker had expressed two years earlier, that someone else would arrive at the theory of natural selection before Darwin published his, had now been realized.

Darwin, aghast, turned to his friends for advice. They suggested a joint presentation on the subject to the Linnaean Society. Wallace agreed, and just over a month later this is precisely what they did. Curiously, the short papers they produced ignited no controversy. The world seemed not to notice. But because of Wallace's parallel experience, Darwin was now forced to produce the long-delayed book. He did so within fifteen months. A mere pamphlet as compared with the mammoth work he had planned, *On the Origin of Species by Means of Natural Selection* ran to 502 pages. It was published on 24 November 1859. The first printing, 1,250 copies, sold out the same day.

Thomas Henry Huxley – Darwin's 'bulldog' – was the foremost advocate of the theory and skilfully demolished Bishop Wilberforce's arguments at the famous meeting of the British Association for the Advancement of Science in 1860.

What Darwin had accomplished was to demonstrate how, through an exceedingly gradual (passive) adaptation to the environment and through changes from generation to generation, a species may diversify or simply become better attuned to its world, producing, ultimately, a creature which is different in form from its ancestor. Thus, as the ages passed some species would remain the same while still others would emerge; and the arbiter for their survival or extinction Darwin called natural selection. Those creatures best fitted to their environment in competition with others survived, while others did not. The picture was one of a steady progression of biological complexity, the most sophisticated product of which is *Homo sapiens*. Such, at any rate, was the inescapable conclusion, though Darwin confined himself to the modest comment that 'much light will be thrown on the origin of Man and his history.' But he must have known that his book would meet a stormy reception, and it did.

The vignette shows Bishop Samuel Wilberforce, Sir Richard Owen's mouthpiece at the British Association meeting. He stoutly denied our relationship to the apes but finally had to concede defeat.

Charles Lyell and Joseph Hooker were, of course, behind Darwin, and so was Thomas Henry Huxley – the best geologist, the best botanist, and the best zoologist in Britain. But there were hostile reactions from Philip Gosse, the devoutly religious father of the novelist Edmund Gosse, from Adam Sedgwick, and from Sir Richard Owen. And so the debate began.

Before publication Darwin had written to Wallace, 'I think I shall avoid the whole subject [of the origin of Man], as so surrounded by prejudices, though I fully admit that it is the highest and most interesting problem for the naturalist.' In contrast with the energetic young man who had returned from the voyage of the *Beagle*, the Darwin of 1859 suffered from chronic lassitude and avoided social contact whenever he could. His condition has been described by some as the result of Chagas' disease, a parasitic ailment which he may have contracted during the voyage, and by others as a psychoneurotic device permitting him to retreat from society and concentrate on his work. At any rate, it was more than six months before the crucial confrontation between Evolutionists and Creationists took place. The occasion was the annual meeting of the British Association for the Advancement of Science, held in Oxford. Darwin was absent. The protagonists in the famous debate of 1860 were Bishop Samuel Wilberforce (who was a mouthpiece for Richard Owen) and Thomas Huxley. The verbal battle between the two followed the presentation of a paper by a Dr Draper, an American, on 'Intellectual Development, Considered with Reference to the Views of Mr Darwin.' In the lecture hall, crowded with some seven hundred students, the atmosphere was tense. The audience must have sensed that a watershed between the age of Creationism and the age of Evolutionism had been reached.

Wilberforce, an outstanding orator, now rose and began an eloquent attack on Darwin's thesis. He had been thoroughly primed by Owen. But in the end, his eagerness to score a point was his undoing. Turning to Huxley, he asked with barbed sarcasm, 'And you, sir, are you related to the ape on your grandfather's or your grandmother's side?' At this Huxley murmured to himself, 'The Lord hath delivered him into mine hands.' He then rose, brilliantly expounded the scientific questions at issue, and only then returned to Wilberforce's clever gibe. 'A man has no reason,' he said, 'to be ashamed of having an ape for a grandfather or a grandmother. If I had the choice of an ancestor, whether it should be an ape, or one who having scholastic education should use his logic to mislead an untutored public, and should treat not with argument but with ridicule the facts and reasoning adduced in support of a grave and serious philosophical question, I would not hesitate for a moment to prefer the ape.' Gales of laughter greeted this riposte, and the humiliated Wilberforce had to concede defeat.

Evolution had won, at least for the moment. For the first time in history it was possible to discuss the animal origin of *Homo sapiens* and its implication for human communities in an atmosphere that was not predominantly hostile. Nevertheless, Darwin waited until 1871 before making explicit his views on where humans fit into the grand scheme of evolution. In *The Descent of Man, and Selection in Relation to Sex*, which in reality consisted of two books in one, Darwin styled humans as descendants of an ape stock, noting the resemblances between apes and humans in physical form, physiology, susceptibility to diseases, and even some psychological characteristics such as instinct, emotions, and sociality. He expanded on some aspects of this latter subject in *The Expression of the Emotions in Man and Animals*, published in 1872.

Perhaps more prophetic than anything else in *The Descent of Man* was its suggestion that the African continent was the cradle of mankind. Darwin reasoned that 'in each great region of the world the living mammals are closely related to the extinct species of the same region. It is, therefore, probable that Africa was formerly inhabited by extinct apes closely allied to the gorilla and chimpanzee; and as these two species are now man's nearest allies, it is somewhat more than probable that our early progenitors lived on the African continent than elsewhere.' The current accumulation of fossil evidence strongly supports Darwin's hunch – whether or not it can ever be finally proved.

Meanwhile, the first pre-human skulls had been unearthed – and rejected as having nothing to do with human evolution. In the summer of 1856 workmen in the Neander valley, a steep-sided gorge not far from the German city of Düsseldorf, had blasted open a small cave some sixty feet above the waters of a tributary to the Rhine. Once inside the cave, as the men hacked through the rubble in their quest for limestone, they came upon some ancient bones. Because the workers' primary interest was in commerce, not antiquity, many of the bones were smashed, leaving only the skull cap and a few fragments from the rest of the skeleton. They belonged to what came to be known as Neanderthal Man, a member of the human family who lived between about thirty thousand and a hundred thousand years ago. (This Neanderthal skull was at least the second of its type – the first had been found in Gibralter in 1848, but was largely ignored.) The reaction to Neanderthal Man, nevertheless, was mainly one of revulsion. The owner of the skull was diagnosed variously as 'brutish,' 'of a savage race,' and 'a pathological idiot.' Professor F. Mayer of Bonn reached the conclusion that the skull and associated bones must have belonged to a Mongolian Cossack on his way through Prussia in 1814 in pursuit of Napoleon's fleeing army; pain from advanced rickets had not only caused the horseman to furrow his forehead, thus explaining the prominent brow ridges, but had also forced him to rest in the cave, where he had then expired – imaginative, but not very scientific. Neanderthal Man was almost certainly a specialized form of early man who slid into extinction.

In 1868 it was once again a party of workmen, this time on a railway line through the cliffs of Les Eyzies in the Vézère Valley in south-western France, who made the discovery. In a rock shelter known as Cro-Magnon they found the remains of five individuals with the unmistakable high-domed cranium and small jaw of modern humans. By late 1912, when a discovery of a peculiar sort received wide publicity, excavation in many parts of Europe began yielding up ancient bones and artifacts. Earlier (between 1891 and 1898), in Java, a young Dutchman, Eugène Dubois, had unearthed the skull cap and upper leg bone, or femur, of a creature that was definitely neither man nor ape, but something in between, which he called *Pithecanthropus erectus* (the name *Pithecanthropus*, meaning simply 'ape man' having been suggested by the German scientist Ernst Heinrich Haeckel in 1886 as a suitable one for a human ancestor). English archeology was in the doldrums. So, when fragments of a skull were discovered in a gravel pit in the south of England towards the end of 1912, the specimen was welcomed into the family of man with almost unseemly haste. One reason was that it had the right credentials: a large skull, whose antiquity was testified to by an ape-like jaw that was unearthed soon after. The 'fossil,' known as the Piltdown Man, appeared to be the ideal 'missing link.'

Forty years and mountains of publications later, Piltdown Man was proved to be a forgery, the ingenious juxtaposition of a modernish human skull and the jaw

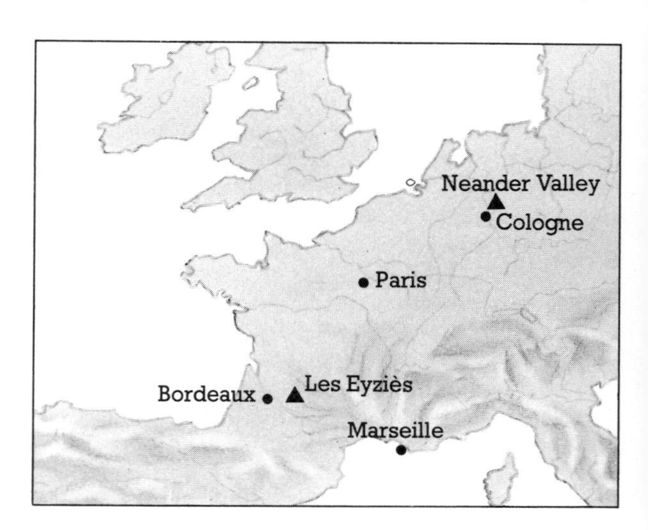

A map showing the locations of the Neander and Vézère sites. Both are in limestone regions.

of an orangutan. In contrast to the reluctance with which Neanderthal Man was greeted, the Piltdown forgery illustrates the sometimes indecent eagerness with which scientists will accept what they want to believe. Researchers today are not exempt from this weakness, and it can be seen in all branches of science. But because theories in archeology are often constructed from relatively little data, in that field the danger of over-interpretation, and therefore biased theories, is particularly acute.

And because of this same weakness, truly important discoveries may be ignored. Such, for many years, was the fate of an article by Raymond Dart, a professor of anatomy at the University of Witwatersrand in South Africa, which appeared in the British scientific journal *Nature* at the beginning of 1925. At a time when much of the scientific world was still seduced by the Piltdown forgery, the discovery of a supposed human fossil in such a far flung and basically barbaric land as South Africa was treated with contempt and scorn. Dart had been interested for some while in fossil deposits that were being quarried at a place called Taung. His searches were rewarded towards the end of 1924 when he found the remains of a skull of a creature that was neither ape nor human. Dart became convinced that the skull had an important place in the stages of human evolution, and he said so in his paper to *Nature*. But it was to be many years before the scientific establishment could bring itself to swing its gaze away from Europe to Africa as it searched for man's origins.

The human tendency to distort the conclusions of science when they cannot be ignored has had some undesirable consequences. One of the most pernicious followed the acceptance of Darwin's theory of natural selection, a version of which was harnessed by thinkers wishing consciously or otherwise, to justify social and economic stratification in industrialized countries. Protagonists of this line of argument, known as Social Darwinism, failed to realize that the social order to which they addressed themselves was the outcome of human cultural capacity, and was not in the natural order of things. Born in the nineteenth century, Social Darwinism has trickled into various social philosophies ever since, manifesting itself in its most horrific form in Nazi Germany.

On the other hand, when in 1975 a Harvard professor, Edward O. Wilson, attempted in a monumental book called *Sociobiology: the New Synthesis* to reach a synthesis concerning the place of mankind in the animal kingdom, he was attacked by radical scientists for introducing what they saw as 'just another, more sophisticated, form of Social Darwinism . . . another attempt to demonstrate that the present social order is natural, inevitable, and unchangeable.'

Passion still runs high when humanity tries to arrive at a true concept of itself, just as it always has.

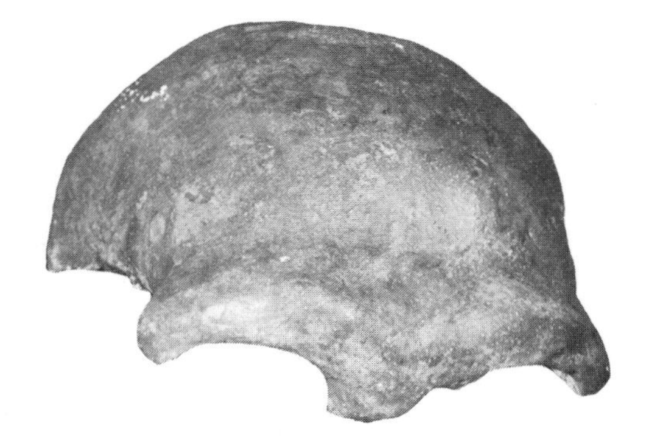

The skull cap of Neanderthal Man – perhaps the most famous fossil find of all time – discovered by some workmen in 1856 during blasting operations in a limestone cave. On the basis of this find many fanciful reconstructions were made, mostly suggesting that the owner was 'brutish' or a 'pathological idiot'. He was neither, but an early Homo sapiens.

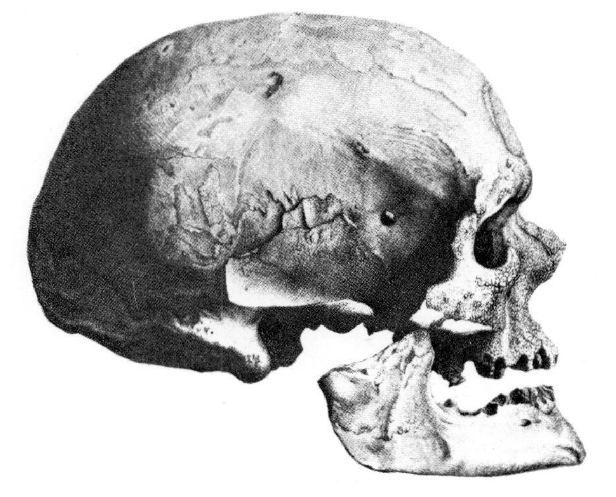

This drawing of a Cro-Magnon fossil skull was made in the 19th century and clearly shows the modern anatomy.

3 The Roots of Humanity

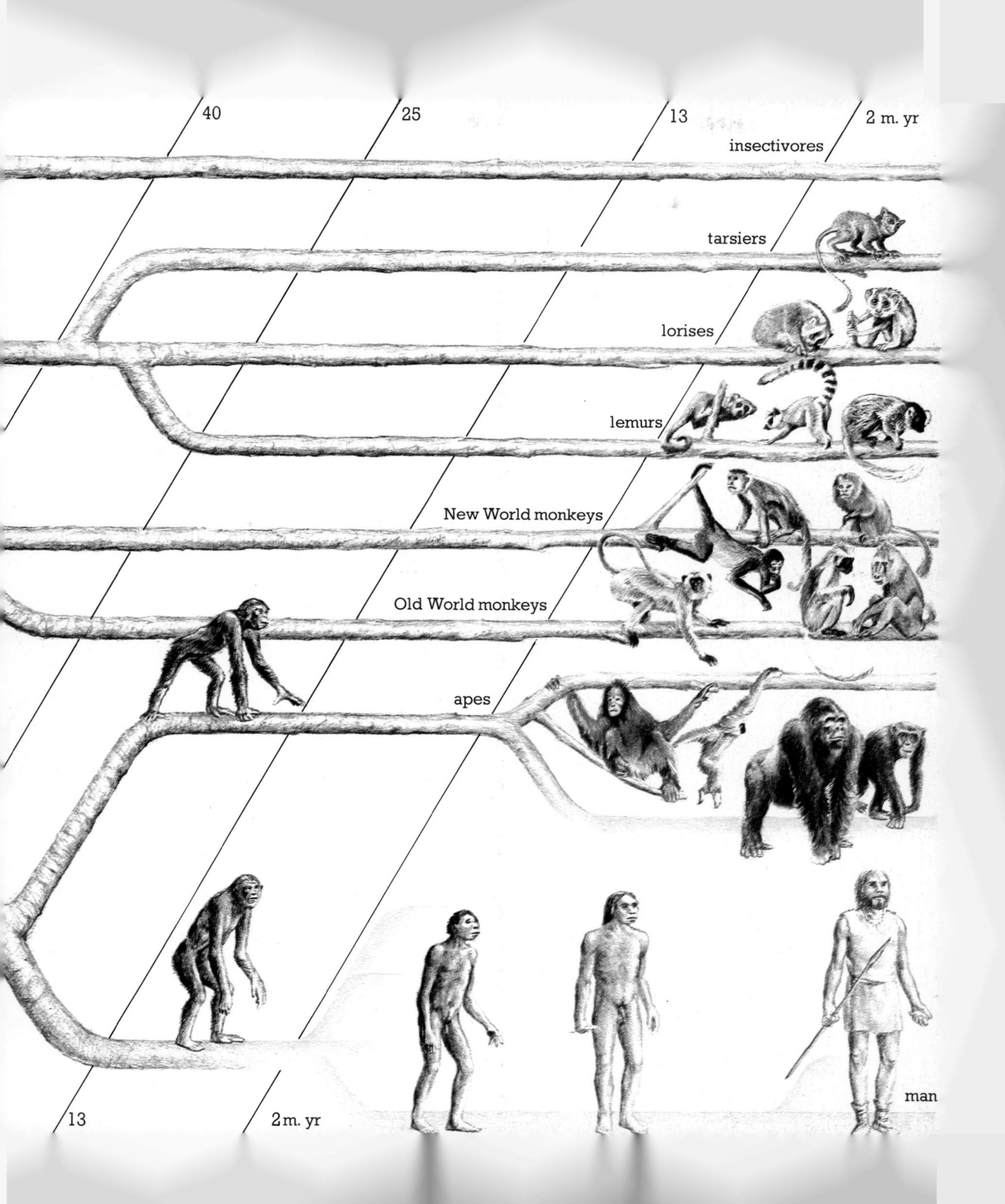
40
25
13
2 m. yr
insectivores
tarsiers
lorises
lemurs
New World monkeys
Old World monkeys
apes
man
13
2m. yr

Around twelve million years ago the earth was in a state of change that was to be crucial for the evolution of man. For perhaps two hundred million years before that time, according to an hypothesis currently accepted by most geologists, the single supercontinent known as Pangea, of which the earth's spherical mantle had originally consisted, had been in the gradual though often violent process of breaking apart into the continents as we now know them. Much the same tectonic movements continue today, making the Persian Gulf a potential ocean, threatening Californians living along the San Andreas fault with catastrophic earthquakes, and incidentally contributing to the discovery of human ancestors buried in the East Africa region of the Great Rift Valley. By twelve million years ago, the buckling of the land had thrown up great mountain ranges – the Himalayas, the Rockies, and the Andes. This upheaval, combined with a steadily dropping world temperature began to bring about a change in the environment without which the evolution of humans would have been impossible.

According to the best guess geologists have been able to make, vast, dense forests carpeted much of Europe, India, Arabia, and East Africa at this time. For almost sixty million years these forests had sheltered the primates, and had nurtured their steady evolution. But, as a result of global environmental changes, the secure green canopy probably began to shrink, and to be partly replaced by relatively open grassland or savanna. The forest therefore exposed its inhabitants to a new ecological pressure: open terrain, which

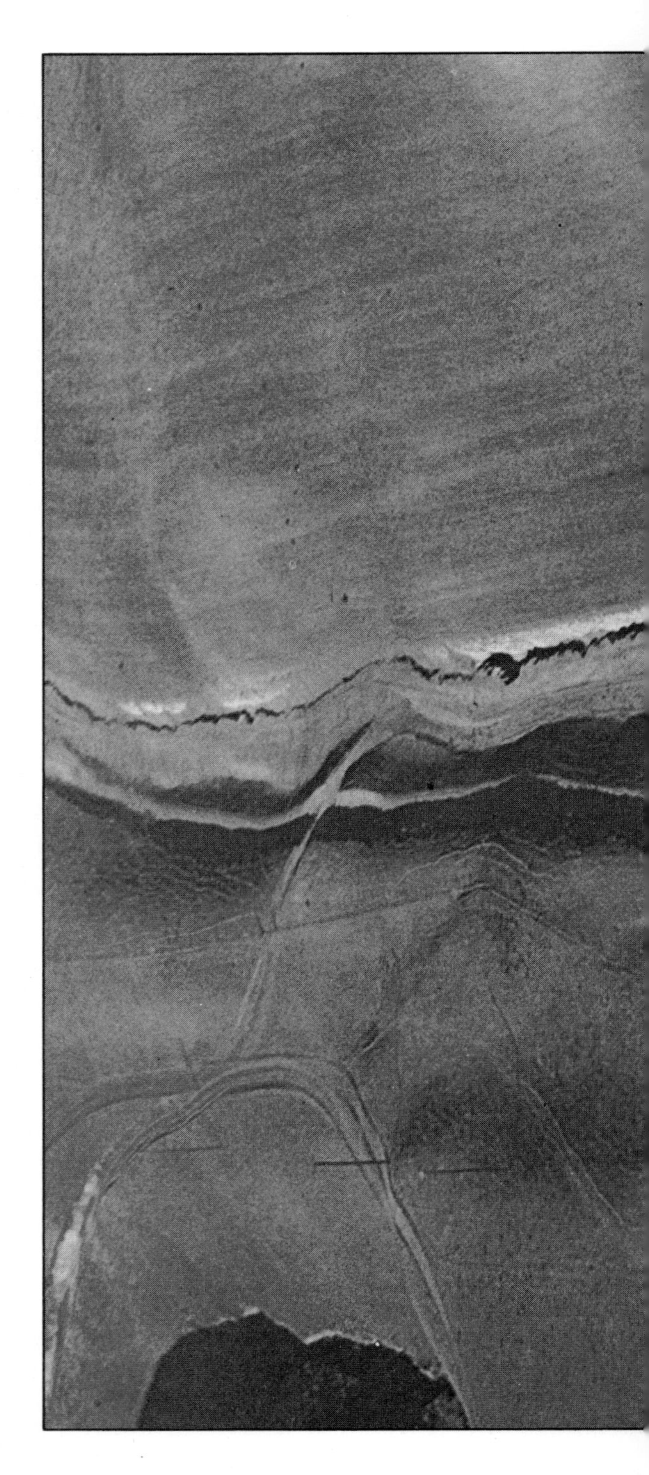

Right: An aerial photograph of part of the San Andreas Fault that carves its way for 700 miles through California north of Los Angeles. This fault system marks the boundary between two gigantic plates of the earth's crust. The Great Rift Valley in East Africa is a similar feature.

Previous pages: This diagram illustrates the relationships of the various living 'families' of the primate order and the geological periods during which it is thought these families evolved. It can be seen, for example, that the primates and insectivores come from a common stock, that the New World monkeys are more primitive than the Old World genera and that several of the Ramapithecus *descendants became extinct. The distribution of the living species of primates, is shown on p.50.*

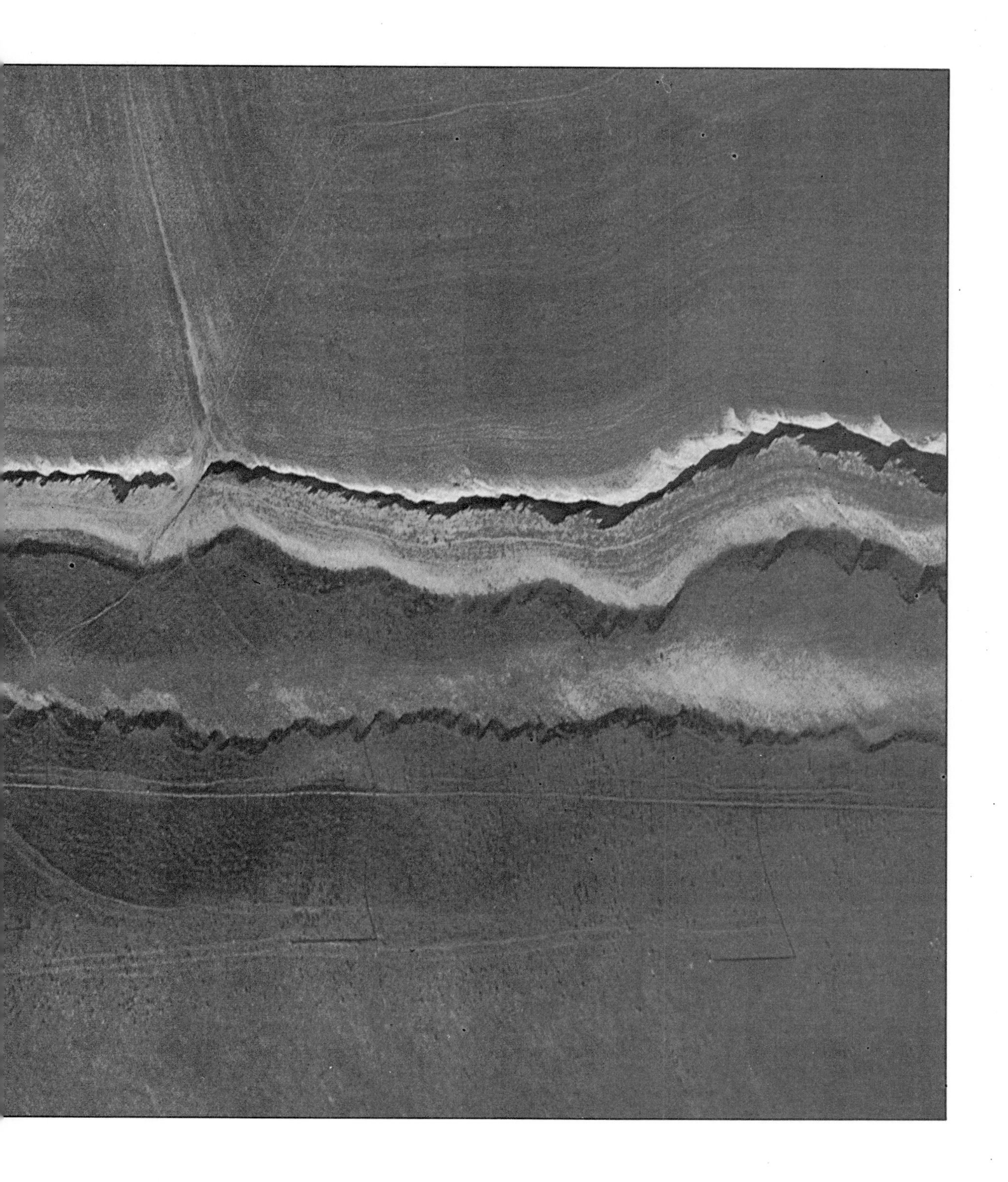

offered a way of life very different from that of an arboreal existence – including predators to which forest dwellers were not normally exposed. With these new demands came a new opportunity as well, and one of the forest dwellers, a small ape, probably around three and a half feet tall, took advantage of it.

Had the climatic alterations come, say, fifteen million years earlier than they actually did, it is arguable that the forest inhabitants would not have had among their number an animal capable of exploiting its new situation in such a way as to become the direct forerunner of *Homo sapiens*.

We cannot escape the fact that we are the product of propitious circumstances molded by the laws of natural selection. The secret of human evolution is extreme adaptability, and the simple physical change that made this possible was the liberation of the hands from the basic function of locomotion. The implications of this modest behavioral change are enormous, for not only does it open the way to technology through the manufacture and manipulation of tools, but it means that the development of language becomes feasible when appropriate selective pressures are operating: a mouth that is adapted to help procure food, carry objects, and threaten or exert aggression, is unlikely to be able to articulate complex sounds. By our definition, competent manipulative hands and a sophisticated language are essential faculties to a cultural animal: the two faculties combine to permit intentional shaping of the environment to chosen patterns. The freeing of hands by the simple expedient of walking upright on the hind limbs was part of an intricate complex of evolutionary behavior, which may have involved such diverse factors as diet, protection against predation, or a change in social organization. This behavioral complex, once initiated, fed back on itself, pushing evolution faster and faster, eventually to produce the human species. The adoption of upright walking could, however, be readily achieved only by an animal that was already in some way pre-adapted to do so, and a life in a forest environment provides the right kind of apprenticeship for two-legged walking.

We can trace our lineage back to an ape-like ancestor that moved dextrously around the trees of its forest home. There are several evolutionary paths that could

The acrobatic silvery gibbon from Borneo demonstrates the arm-swinging (brachiation) locomotion that is typical of this genus.

be followed from this point so far as types of locomotion are concerned and indeed many of them have been explored. For instance, the gibbon has won for itself the accolade of being the most accomplished acrobat in the world. But it has paid the price of having to remain relatively small (weighing around twenty pounds) and to be equipped with a hand that, though superb at split-second swinging through branches, is poor at manipulating objects. The orangutan followed a similar path, but, being substantially bigger, is much less acrobatic. In contrast, the gorilla and chimpanzee both left the trees to a greater or lesser degree, but evolved a specialized form of locomotion of their own, called knuckle walking. Only man and his ancestors habitually moved around in an upright position.

The period during which savanna began to replace the forests was therefore crucial in the evolution of upright walking, and thus of humans: if it had come earlier, today's primates might have been dominated by monkeys running on four legs, like baboons; if it had come later, the apes might have all been channeled along the path of the gibbons. Either way, the eventual arrival of *Homo sapiens* would have been greatly delayed, if it had ever taken place at all.

Before exploring the reasons why the behavioral complex that created the human family from an apelike stock evolved the way it did, we should take a look at the long-extinct ape that eventually ventured from the forest fringe to live in the open. Since humans owe their primacy on land to a long evolution in the trees, we should know from where it acquired the physical equipment that enabled it to survive and thrive in the challenge of the new environment. And to demonstrate the powerful influence of environment in shaping evolution, not only in physical form but also in social organization, it is instructive to explore the social life of some of our closest primate cousins.

The beginning of the Age of Mammals, some seventy-five million years ago, had been marked by geological stirrings and climatic instability that favored creatures able to maintain easily a constant body temperature. The small mammals that for almost a hundred million years had been snuffling unobtrusively about the forest floor in search of seeds and insects, in constant danger from reptilian predators, were offered the new evolutionary opportunities. We are primarily interested just now in that small, long-snouted bewhiskered creature, rather like the tree shrew of today, which about 70 million years

A reconstruction of Plesiadapis, *a fossil prosimian, the remains of which have been found in the Paleocene formation of Europe.*

The specialized form of quadrupedal locomotion known as knuckle walking is used by the gorilla and, as shown here, by the chimpanzee.

A baboon using its hands for holding food while feeding.

ago abandoned ground living and took to the trees.

The two main primate characteristics imposed by life among the branches of trees are grasping hands, with flat nails rather than claws, and a superior eyesight with stereoscopic vision (eyes at the front rather than the side of the head). Together with an ability to hold the body in a vertical position, either during feeding or moving, these features form the true primate package. Other mammals have part of it (for instance, most predators have stereoscopic vision) but only primates have it all. And when ancestral humans returned to ground living they had no need to acquire new abilities: they simply improved on the old ones.

When the first rat-like primate scurried up the trees to escape from growing competition and predation on the ground it was embarking on a perilous adventure. Instead of the relatively secure foothold normally experienced in life on the ground, tree-dwellers have to contend with never-ending hazards of unpredictable twigs and branches and the surrounding cloud of foliage: a twig may twist and turn, or break off suddenly,

Chimpanzees' hands are capable of fine manipulation. Here a mother is holding a stick with which she had been 'fishing' for termites.

having been snapped by some previous passerby; the danger of breaking twigs is constant, so this must be monitored, and reactions must be swift and sure when needed. The keen selection pressures of this new ecological niche must have produced a basic stock that was superbly adapted to the new environment, for we have the highly successful primate order to prove it.

Even the accomplished acrobats of today, the gibbons, apparently still come to grief occasionally, probably more often when a rotten branch gives way under their relatively modest weight than as a result of misjudging distance. The exigencies of life aloft must have given natural selection a particularly keen cutting edge, speeding along the process of evolving sophisticated mechanisms, such as an opposable thumb, for coping with the challenges of everyday life.

Grasping hands and front-set eyes for stereoscopic vision may, however, be only part of the primate picture. If the early prosimians had been vegetarians instead of insect-eaters the primates might not have evolved in the way they did. Some people argue that the pressures of stalking insect prey along slender branches, followed by a cautious and accurate grasping of the prey, was at work in the evolution of primates as we know them.

Whatever the main evolutionary pressures may have been, we know that the eyes and the hands underwent marked changes. From a clawed paw in the earliest prosimians, there developed a hand with separately movable fingers. The most sophisticated advantage of this feature is, of course, the development of what is known as the precision grip, in which the thumb and index finger meet at their tips to form a circle. Humans use this grip for very fine manipulative maneuvers – and we also use it, significantly, as a gesture to indicate perfection!

The ability to form the precision grip depends on a number of properties of the hand, not least of which is the freedom to move the thumb in a swinging action across the palm of the hand. This facility, known as opposability of the thumb, makes only a gradual appearance through primate evolution, and even though many monkeys and apes are capable of some kind of precision grip, only in humans is it fully developed. For instance, the monkeys of the New World tend to bring the index finger and thumb together in a kind of scissors action. Old World monkeys do a little better, some achieving a very respectable opposition. It should come as no surprise that the monkeys with the best grip are those which need it for survival; matching body structure with ecology and behavior is inescapable in the context of evolutionary biology.

The gelada baboon, a ground-dweller in the high plateaux of Ethiopia, feeds on seeds, grasses, bulbs and insects which it picks up from the ground as it shuffles around on its haunches. Unlike most monkeys, the gelada has a large thumb relative to its index

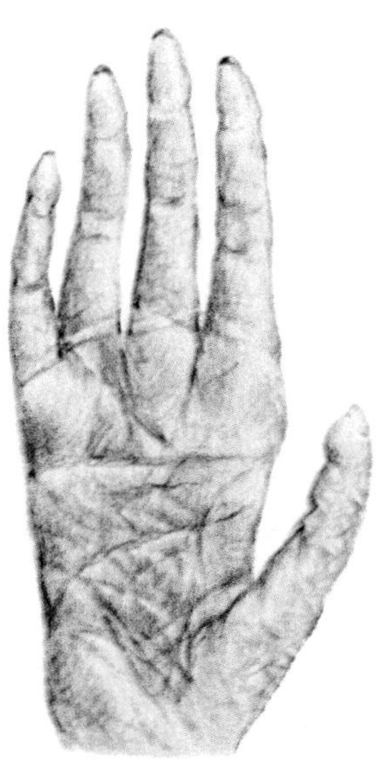

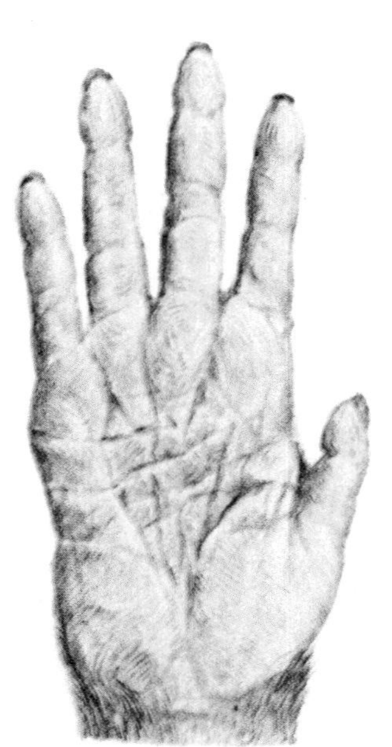

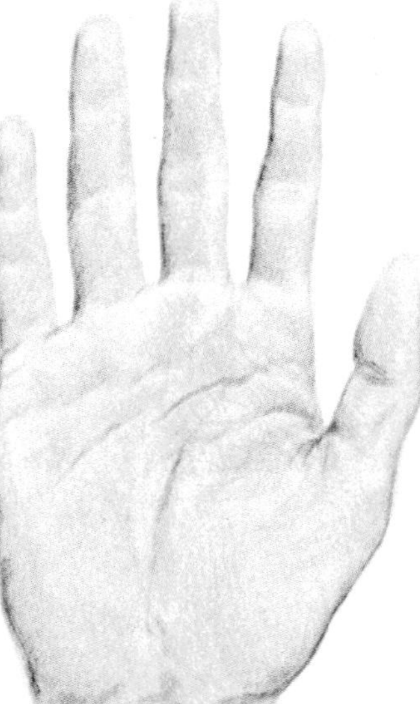

The evolution of our hands and the manipulative skills that have been developed have been important factors in the development of our brains and vice versa (see feedback on p.73). On these two pages we compare (below left) the hands of a gibbon and chimpanzee (apes) and Homo. *The precision grip of the chimpanzee (left) although capable of fine manipulation, is not as sophisticated as ours. The development of the power grip preceded the precision grip and is illustrated on the right – in the human hand – and below we show the culmination of the precision grip, which allows the fine manipulation of tools.*

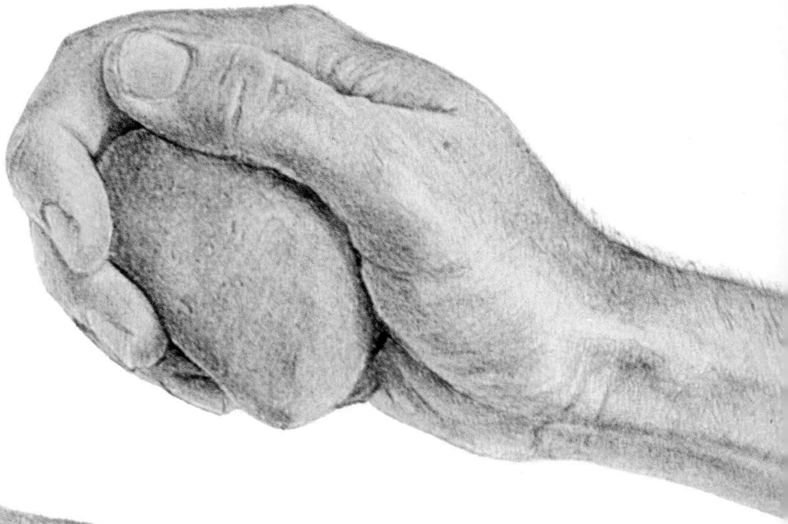

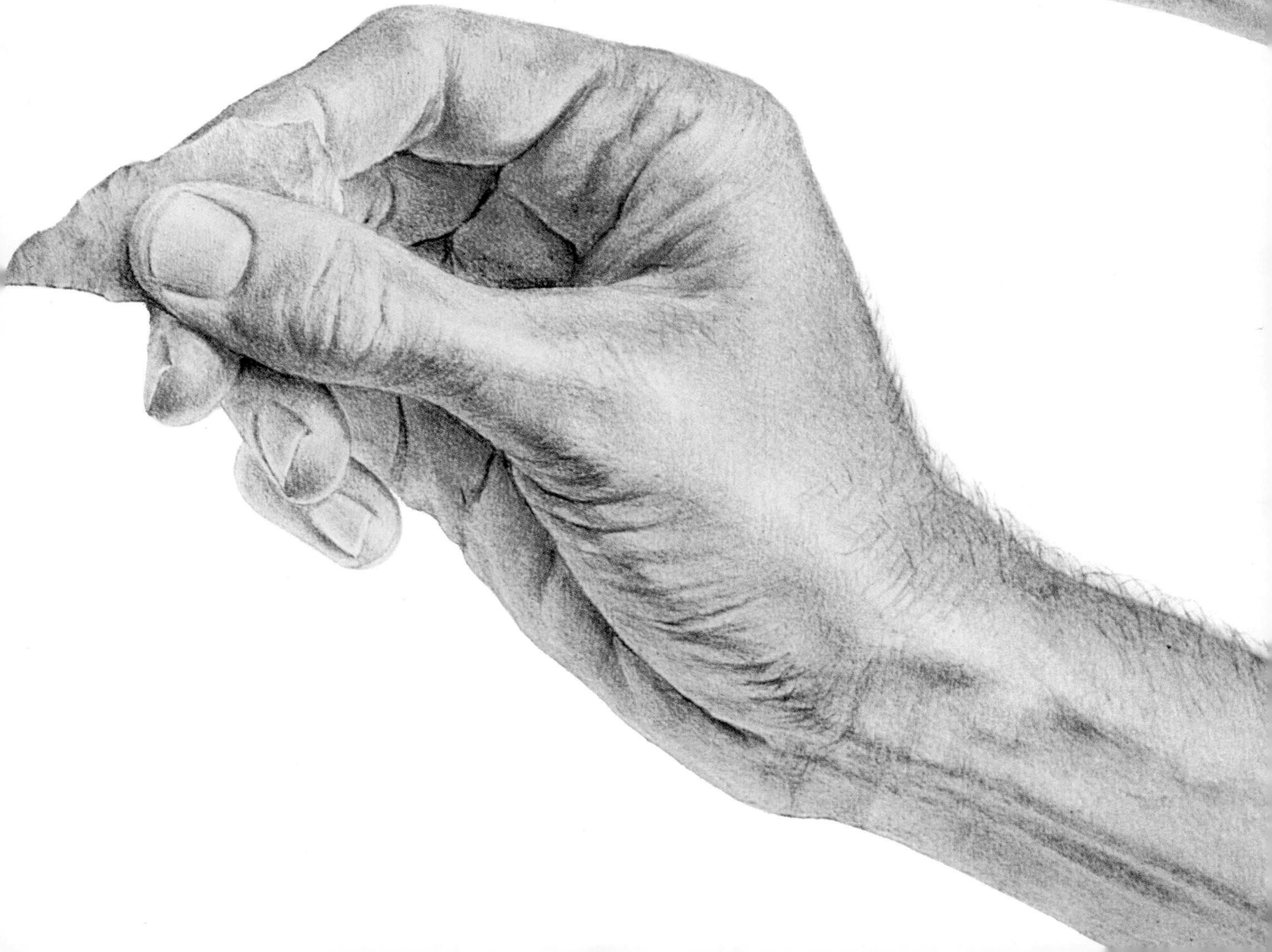

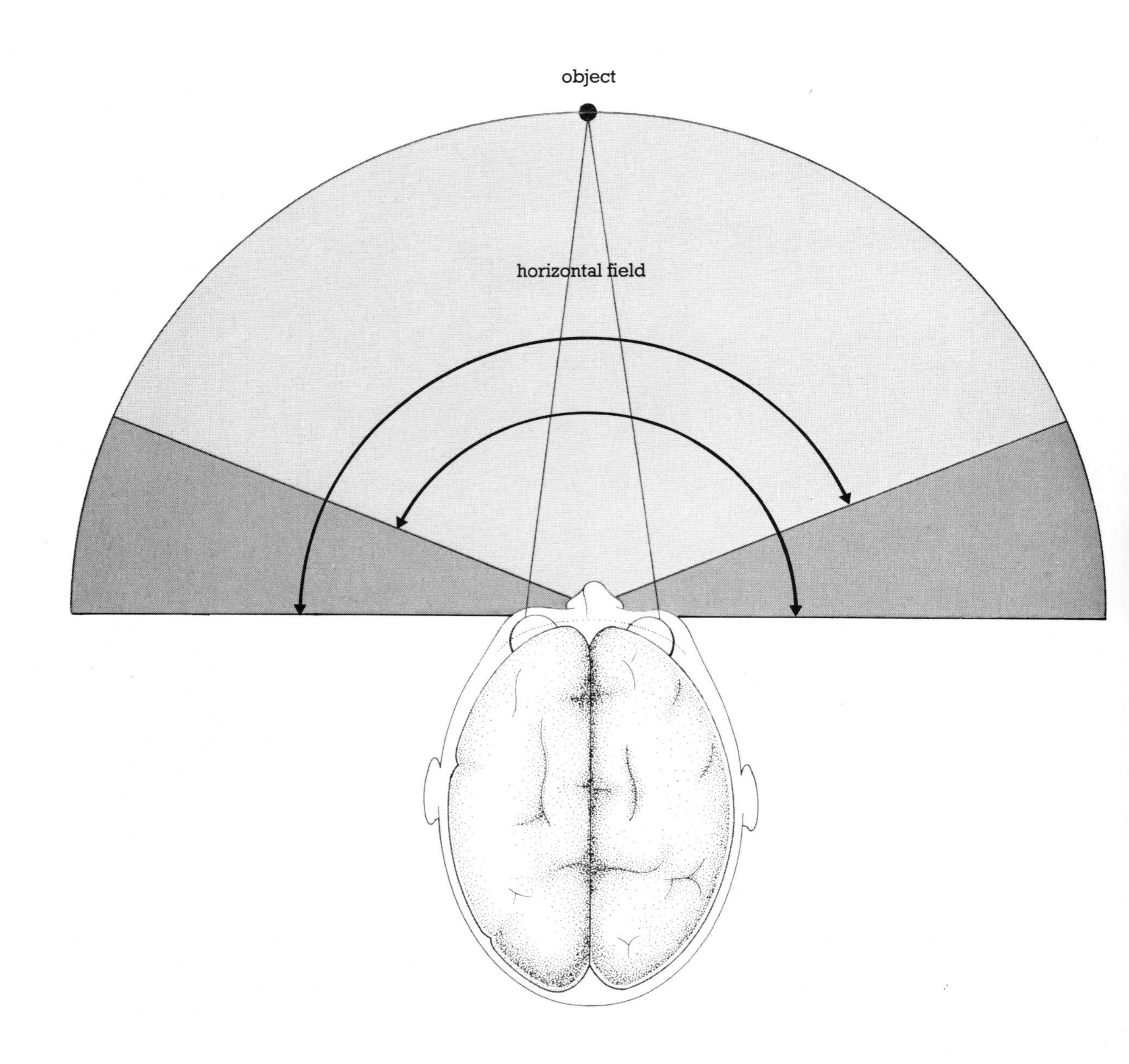
object
horizontal field

finger (this finger has in fact shrunk somewhat), thus permitting a better precision grip for its food-gathering. By contrast, the fruit-eating, acrobatic gibbon has practically no thumb at all, the appendage being something of a liability in the ape's intrepid aerial displays. Its fingers are especially attenuated as well, making a precision grip impossible. But as an arboreal fruit-eater, the gibbon has little need for fine manipulation conferred by precision grip. It would be fair to say that a primate with gibbon-like mobility which fed on small objects is at least biologically improbable.

Chimpanzees, by human standards the most intelligent of all the non-human primates, can achieve a precision grip of sorts, but because of the adaption of the chimpanzee's hand to knuckle walking, it isn't as good as in humans. The chimpanzee brings its thumb to the side of its bent index finger, a position in which it can perform some remarkably fine manipulation, as it demonstrates by its ability to strip a twig of its leaves and then 'fish' with the twig through the narrow entrance to a termite nest.

Overall, then, the evolution of the hand through the primate order brought an increasing ability to grasp, important both for moving safely on relatively thin branches, and for collecting and holding food. Later in the evolutionary progression the precision grip began to emerge, reaching extreme sophistication in humans. But, in spite of this specialization, the human hand has not lost its basic general abilities, such as the power grip and prehensibility. This is yet another demonstration of the key to success in human evolution: we have developed specializations, but not, as in the gibbon, at the expense of general abilities – the human hand is functionally extremely flexible.

Fine manipulation with the fingers has implications beyond simply being able to pick up small food morsels. Together with that other major characteristic of primates – stereoscopic vision – it transforms the animal's appreciation of its world. Because many of the earliest primates were nocturnal, the size of the eyes likewise increased. Later, as the larger primates became more active during the day, the nerve network at the back of their eyes became even more sophisticated, further refining their discrimination of light and dark to the perception of color.

The combination of stereoscopic vision and the ability to see the world in color (see photographs on next page) were important evolutionary factors, allowing the higher primates to become more aware of their environment. The diagram shows the total horizontal scan of our eyes: the central area where the fields of vision overlap give us binocular vision, while the left and right areas indicate the outer fringes of vision.

For those of us who are totally accustomed to color vision, it is worth pausing to consider the tremendous benefits that this perception confers in discriminating objects against a monochromatic background, and as an aid in perceiving depth and distance. In a forest environment made up of kaleidoscopic shades of green, with occasional splashes of color, the usefulness of monochrome vision is severely limited, as the pictures demonstrate dramatically. Even in the single, but clearly important, activity of searching for fruit, being able to perceive color is an enormous advantage. Given enough time, the evolution of color vision in the forest-living primates was inevitable. Once again, when our ancestors returned to the ground they took this with them, giving them a clear advantage over the many ground-livers with whom they would soon be in competition.

For the higher primates, with their ability to see that world in color and in three dimensions, and, also to pick up and manipulate objects, the world becomes more than just a three-dimensional colored pattern. It is also a world full of identifiable objects. Monkeys, apes, and human beings are almost the only creatures that fiddle with things, picking them up, turning them over, inspecting them visually, by smell and by touch.

In doing so, the shape, texture, weight, smell, and utility of an object can be assessed, and that object comes to mean something in the animal's world. It is not merely a small part of a larger intact pattern; it is one object separable from among many others. The implications of this new mental dimension are vast; indeed they become essential as the background against which the development of the human mind is to be understood. The opportunities for learning about the world, rather than simply reacting to particular shapes in a pre-programmed fashion, are enhanced enormously. And, ultimately, the ability to view objects as separate entities is an absolute prerequisite for the evolution of language, which is possibly the one unique human attribute. In a very real sense we owe our capacity for speech to the higher primates' reaching out to analyse their three-dimensional world.

The photographs show how much easier it is to see objects – food, for example – in color than in monochrome.

Right: Fiddling with things, as this chimpanzee is doing, is an extremely useful way of exploring the environment and assessing the utility of an object.

The living primates vary greatly in size, from the diminutive mouse lemur to the massive gorilla. As far as one can tell from the extremely limited fossil evidence, some of the living primates are very similar to their extinct ancestors, so that by looking at contemporary primates we can obtain at least a glimpse of the stages of evolution through which the human ancestor passed. For if a species is well adapted to a particular ecological niche, the pressures for change will pass it by. The species will remain roughly the same for a very long time. This, apparently, is what has happened with many of the primates.

Further along the evolutionary path, however, the exercise becomes less useful. We cannot look at chimpanzees and gorillas, our closest primate relatives, and declare, 'Ah, this is what we once looked like.' The apes that we see today are the result of evolution over the past 15 million years, just as much as we are. Our ancestor was much smaller than contemporary humans, and so too was the ancestor of the gorilla and chimpanzee. On the other hand, just as upright walking is a special human adaptation, so is knuckle walking in the two apes. This perspective should always be borne in mind when apes and monkeys are being studied in the hope of learning about human evolution, not least because the very human-like qualities in some of them, especially the chimpanzees, can seduce even the most hard-bitten scientists into dangerous anthropomorphizing. Nevertheless, animals with which we share close ancestry and which now occupy environments similar to those on which our ancestors lived can teach us something about our own evolution. And it is because of the importance of the environment in shaping evolution of behavior that we can probably learn more about ourselves, or rather our ancestors, by studying the baboon (a monkey), rather than the gibbon (an ape), even though we are more ape than monkey.

Contemporary primates are divided into two main groups or suborders: the prosimians, and the monkeys and apes. The prosimians are small creatures and include the mouse lemur, sifaka, tarsier and slow loris. The monkeys and apes are further divided – into three so-called super-families: the New World monkeys (ceboids), such as the tamarin, howler monkey, and spider monkey; the Old World monkeys (cercopithecoids), including the baboon, patas monkey and colobus monkey; and the apes, including the gibbon, siamang, orangutan, chimpanzee, gorilla and man.

The monkeys of the New and Old World provide an interesting example of what is known as parallel evolution. By this is meant the evolution of very similar characteristics in totally separate species through being subjected to separate but similar environmental pressures. The New World monkeys are limited to Central and South America, whereas those of the Old World are found throughout the tropics of Africa and Asia. Physically the two groups are very similar, and it takes a keen eye to spot the few distinguishing characteristics. One of these is the nostrils: close set in the Old World species, they are splayed apart in American monkeys. Another is the prehensile character of the tail in many New World monkeys. One very clear distinction between the two groups is in the teeth: Old World monkeys and apes have a total of thirty two whereas their American cousins are equipped with a more 'primitive' set of thirty six, with three premolars instead of two in each quarter segment of the mouth.

As it happens, because teeth are particularly resistant to decomposition and to being themselves chewed up and eaten, they have a disproportionately large role to play among surviving clues to the past. (They do, of course, tell a good deal about their former owner's habits, especially diet.) Nowhere is this more true than in the description of what is thought to be the first primate; until recently all that had been found was a single upper molar sifted out of tons of chalky rubble from a site in Montana known as Purgatory Hill. Paleontologists are used to founding bold theories on fragile facts, but that one tooth stretched interpretation to the limit. Since North America has no primates today, how did it get there?

The answer is to be found in the geological hypothesis of Pangea – the single land mass which as it first began to fragment is pictured as having formed two supercontinents, one in the nothern hemisphere (Laurasia) and the other in the south (Gondwanaland). At some point what was to become South America split away from Gondwanaland, and began a gradual drift westward, eventually to be joined with North America a mere two million years ago. Just when South America began moving across the ocean as an island is difficult to pin down geologically, but evidence from fossils suggests that it must have been after the basic primate stock had emerged, some 70 million years ago. It seems likely, therefore, that the land mass that was to become South America had on board a prosimian stock similar to the one left behind in the Old World as the island began its long migration westward.

The presence in Montana of that lone molar suggests that some of the earliest primates also once lived in what is now North America, during a period when that land mass was closely associated with the hypothetical northern supercontinent of Laurasia. The climate then, however, must have been subtropical or tropical in order for that mammal to have had vegetation to eat all the year round. But gradually Laurasia fragmented; there was a slow northward drift and a cooling of the climate. By some fifty-five million years ago, the North American climate had become too cold for primates to survive, and the chances that any could have migrated from elsewhere would also have been impossible. The ancient link with Eurasia was already broken, the eastern land bridge from Asia was too cold for primates to use and the South American land mass had not yet drifted westward.

Things were different in Europe and Asia, however, where as Laurasia drifted northwards, the primates could migrate southwards in pursuit of the disappearing sun. It was not because they were heliophiles that they migrated, but rather because many of them had by now left behind their insectivorous dietary habits and had taken to eating fruit and leaves. In the tropics these are available year round, but in cooler climates the supply is seasonal, as too is an insect diet. So, as the continents migrated north the primates went south, thus remaining more or less stationary with regard to climate and the consequent availability of food.

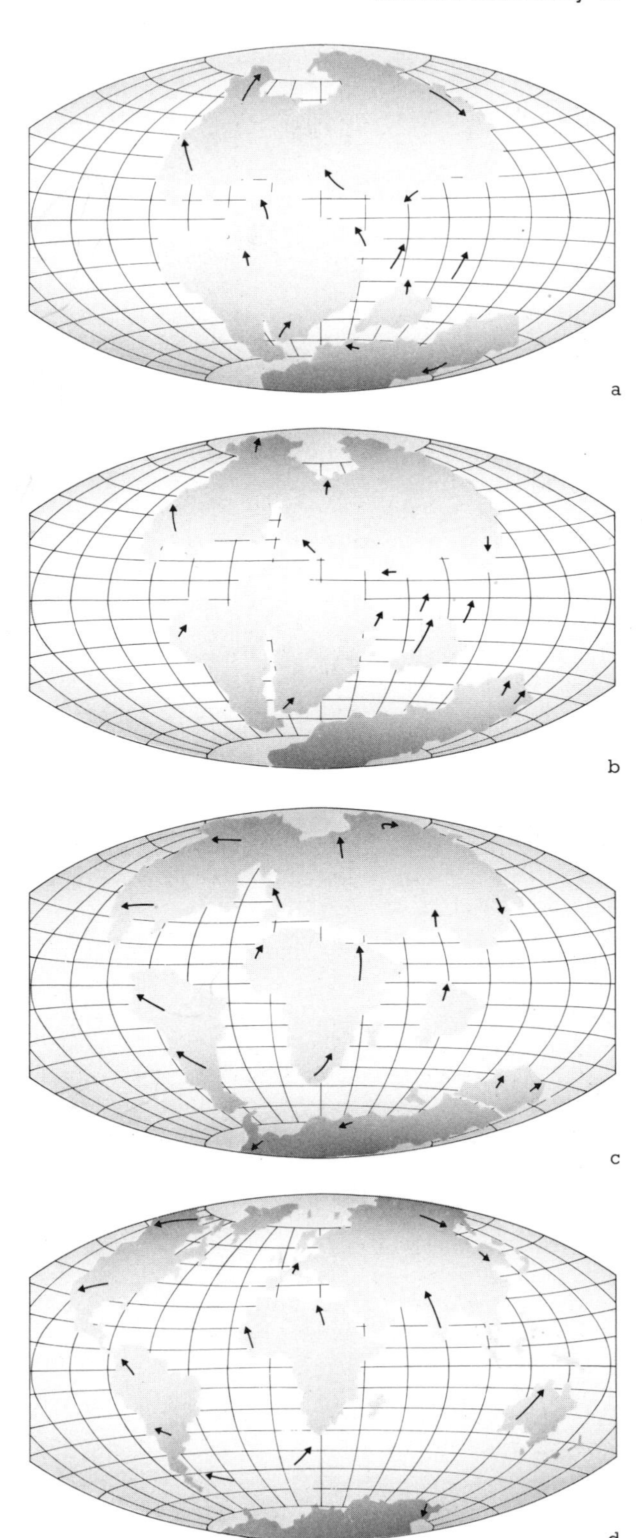

The original single land mass that we now call Pangea began to break up about 200 million years ago (a) into the two supercontinents of Laurasia in the north and Gondwanaland in the south. By about 40 million years ago (b) South America was splitting from Africa, North America from Eurasia and India was drifing northwards. As North and South America drifted further west the Atlantic Ocean opened up, Africa began to collide with Europe, Australia became detached from Antarctica and India was closing on Asia. This was the situation about 65 million years ago (c). By today Africa and India have collided with Europe and Asia, causing the uplift of the Alpine-Himalayan mountain system, South and North America are in contact and Australia is well away from Antarctica (d). The drifting process is still going on: Africa, for example, is splitting apart along the Great Rift Valley.

The study of living primates is important in the search for our ancestors: it helps us to understand the meaning of primate fossils, and their social organization and behavioral patterns may point to similar traits in the early hominids. On these two pages we show the natural distribution of the living primates (other than mankind), and a few examples of the many species in the order. The lesser mouse lemur, the slow loris and tarsier are prosimians; the squirrel monkey and spider monkey are from the New World and show features such as a prehensile tail and divergent nostrils, which distinguished them from the Old World Monkeys, such as the Patas.

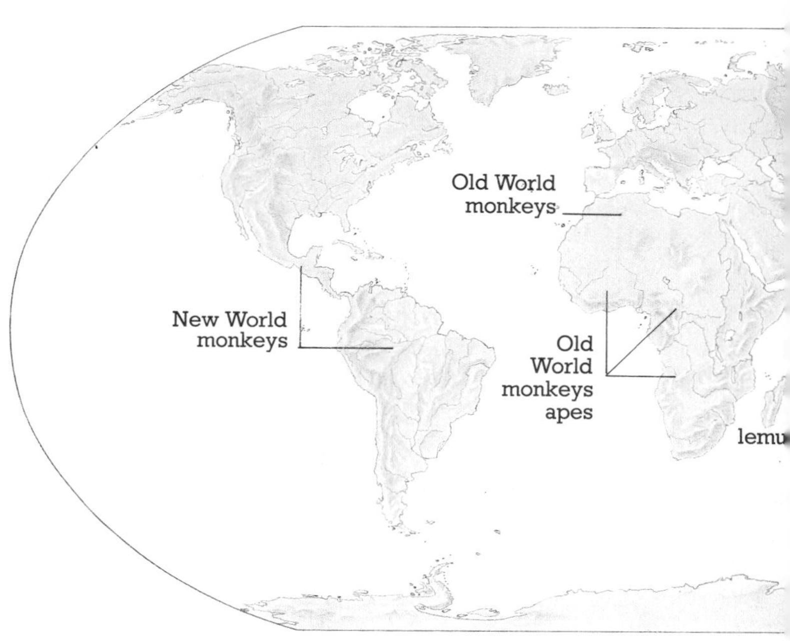

Slow loris

Tarsier

Lesser mouse lemur

Gorilla – silverback male

Squirrel monkey

Patas monkey

Spider monkey

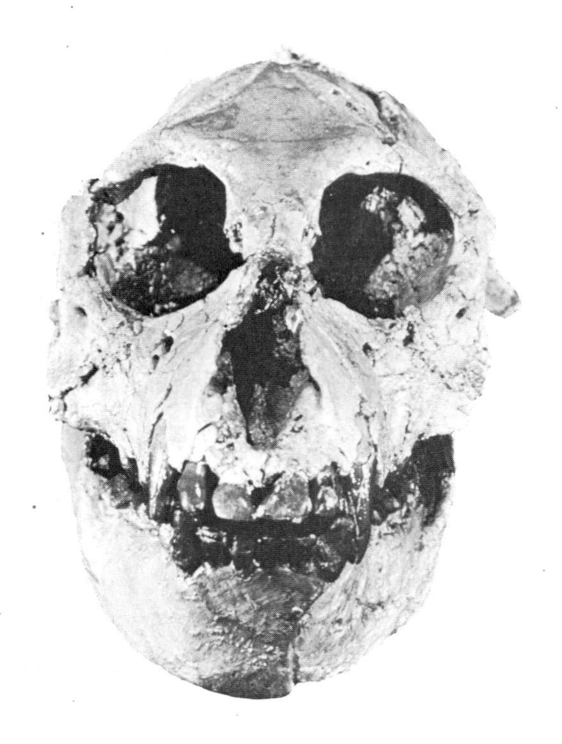

The rather monkey-like skull of Aegyptopithecus zeuxis *from the Fayum Depression – the first ape to emerge from the Old World monkey stock. Some elements are partially restored.*

In South America prosimians seem to have evolved in much the same way as the Old World primates. The common factor, of course, was a life in the trees, an example of the importance of a particular ecological niche in nurturing specific evolutionary traits.

There must have been at least one important difference between the ecological nurseries of the New and Old Worlds, however, and this may possibly have centered around the density of the forest. On its slow journey westwards, South America experienced a relatively stable subtropical/tropical climate, whose conditions favored thick, dense tropical forests. By contrast, the Old World primates were subjected to a slightly cooler climate and more broken forest. Although we cannot be certain, we can at least speculate that this difference is in some way responsible for the emergence of apes in Africa and Asia, whereas they remained absent from the Americas.

Among the Old World primates, there was an added stimulus to southward migration – the steady drop in world temperature that began almost imperceptibly perhaps sixty million years ago, gathering momentum until, by forty million years ago, it was falling sharply. It is no coincidence that this period is marked by an important shift in the evolution of primates.

For more than thirty million years since primates first ventured into their new environment, the prosimians had flourished. From their small rat-like beginnings they had developed front-set eyes, clasping hands and, many of them at least, a dietary shift from insects to fruit and leaves. They were now bigger than they had been, though probably still smaller than a cat, and had established themselves over a wide area. But around thirty or forty million years ago, these prosimians slid into a decline. The reason, almost certainly, was competition from their own issue – the monkeys had arrived. Shortly afterward the apes made their first appearance as well.

We owe much of our direct knowledge of these earliest monkeys and apes to excavations in the Fayum Depression, a wasteland on the eastern edge of the Egyptian Sahara. Arid though it is now, forty million years ago the region was covered by a thick tropical forest through which rivers and streams flowed sluggishly to the Mediterranean. Such conditions are not especially conducive to preserving bones long enough to allow the slow materialization that will eventually transform them into fossils. Some did survive, however, and the Fayum Depression has attracted the attentions of archeologists and paleontologists ever since the beginning of the century. It wasn't until 1960, when Elwyn Simons of Yale University initiated intensive searches, however, that the area began to yield its treasures in quantity.

One of the prize specimens recovered from the Fayum is an ancient ape with the scientific name *Aegyptopithecus zeuxis* (*pithecus*, the Greek word for ape, crops up repeatedly in the nomenclature of human evolution). The Simons expedition was lucky enough to discover a virtually complete skull (except for the jaw), a rare event in the pursuit of human origins; and with an age of around twenty-eight million years it represents the first ape to emerge from the basic stock of Old World monkeys. So, in our search for the origins of our species, *Aegyptopithecus* is an ancestor which we share with living apes.

The double transition in primate evolution that first brought the prosimians to monkey status, and then monkeys to apes, involves identifiable changes in the patterns of locomotion. To a large degree this was influenced by the size of the animals. The small prosimians engage in vertical clinging and leaping – that is, when at rest they tend to cling to vertical branches, thus favoring the upright posture that comes to be characteristic of primates. Their hind legs, in continual use as they leap from branch to branch, are especially powerful. The smaller front legs are important in food-gathering.

Monkeys, which are considerably larger than their prosimian forebears, run along branches on all fours – quadrupedally in other words. Because they are large compared with the size of the branches that support them, the monkeys have developed hands and feet that can fully grasp the supports, unlike most of the prosimians. The tail is useful in maintaining balance as the monkey scurries adeptly through the branches. Modern apes, which are tailless, have evolved a number of specialized forms of locomotion, such as knuckle walking in gorillas and chimps, and brachiation (using the hands to swing from branch to branch or a modified form of this) in the gibbons, siamangs, and orangutan. Their ancestors, however, probably moved about some of the time by swinging beneath the branches, using their hands. By suspending itself from a branch an ape, such as the orangutan which has a larger and therefore heavier body than a monkey, can reach ripe food at the ends of branches: the ape can spread its weight by hanging on with three hands, and plucking fruit with the fourth (it is not incorrect to talk of four hands when discussing the orangutan, because that is just what all four extremities are).

These classifications of locomotion inevitably are generalized, especially when applied to living species. For instance, many modern monkeys in the Old World move about in the ancestral ape way, and it is significant that these are among the larger species – another example of the link between body size and mode of locomotion. Baboons have largely left the trees in favor of terrestrial running and walking on all fours. But the general progression from vertical clinging and leaping, to quadrupedal running along branches, to arm swinging, does hold through the evolution of primates. Locomotion therefore provides helpful clues for deciding where along the evolutionary path fossil primates should be placed.

The slender evidence available from the *Aegyptopithecus* skeleton, suggests that this ape probably

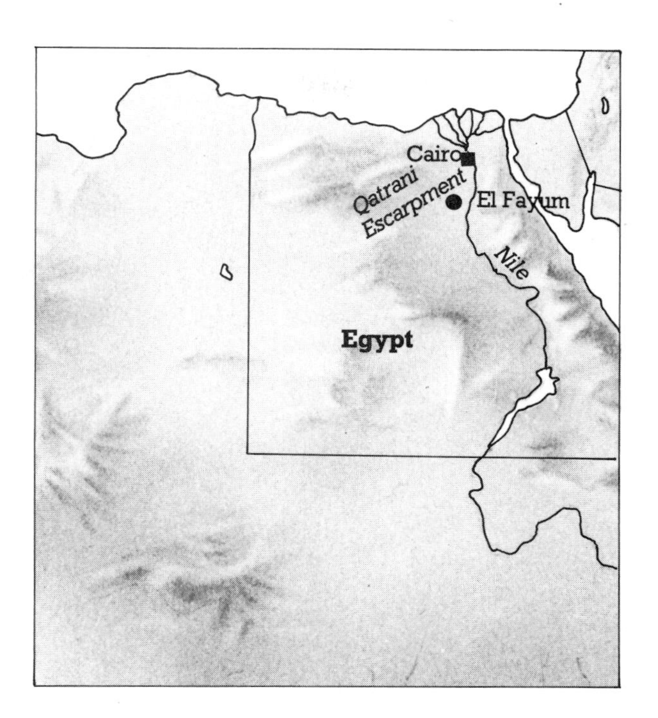

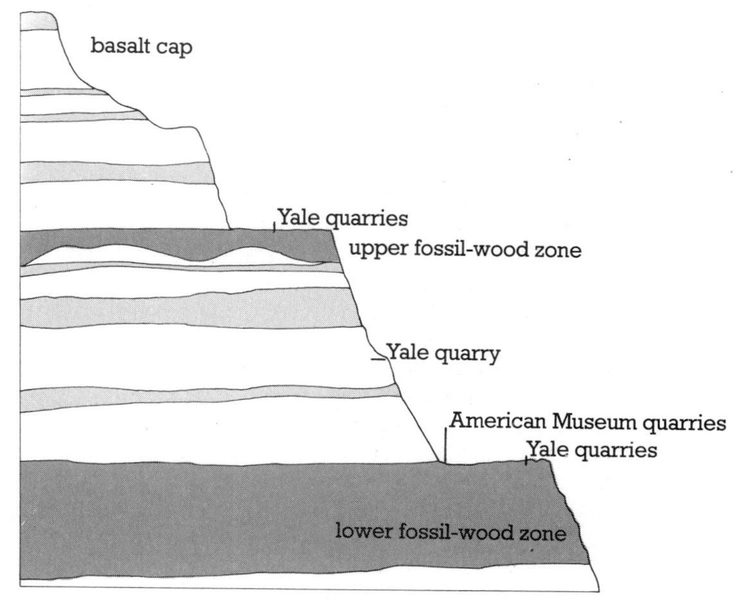

In Oligocene times, the Fayum Depression in the Egyptian Sahara was covered with forests through which rivers flowed north to the Mediterranean.

A possible explanation of the emergence of apes in Africa and Asia and not in the Americas is the more open nature of the tree cover in the former. The photographs (above and left) compare the savanna (in the rainy season) in Tanzania with the Brazilian rain forest.

moved about quadrupedally. Almost certainly, though, it was beginning to develop arm-swinging as a more efficient method of reaching the fruit at the ends of branches, while many monkeys concentrated more on the leaves. This general difference in diet must have been one of the factors in the evolutionary divergence of monkeys and apes.

As we move farther along the path of evolution towards humans the going becomes distinctly uncertain, again owing to the paucity of fossil evidence. The next most significant fossil ape, called *Dryopithecus africanus*, comes roughly eight million years later than *Aegyptopithecus*. It was Louis and Mary Leakey who recovered a nearly-complete skull and some limb bones of this species, excavated in 1948 from Rusinga Island in Lake Victoria. At the time an enthusiastic press hailed the find as the direct common ancestor of humans and apes. Although there is still some dispute as to where this species stands on the evolutionary path, it seems more likely that *Dryopithecus africanus* itself had already diverged from the common stock and was on its way to forming a lineage of apes. However, earlier members of the genus *Dryopithecus*

This orangutan, above left, spreads its weight by using all four limbs. Unlike monkeys the apes do not have a tail to help them maintain balance. The gibbons shown above provide another example and one of them is clutching food with three of its limbs while using its fourth for hanging.

(once called *Proconsul*) may well have given rise to the ancestors of both the human and the ape lines.

The earlier name *Proconsul* deserves mention. It was coined by Arthur Hopwood of the Natural History Museum in London because at the time there was a famous chimpanzee called Consul, and he thought this fossil ape might be an ancestor of chimpanzees. This is one illustration of the way in which scientific names for fossil species are invented, and although it shows the science to have a sense of humor, it has caused problems too. The naming game in paleoanthropology must have caused more battles between experts than almost anything else.

If we now trace the steps of evolution onwards from a dryopithecine (ancestral to both apes and humans) through a period leading up to about twelve million years ago, we find three major genera: *Dryopithecus*, *Gigantopithecus*, and *Ramapithecus*. Once again, there is incomplete agreement about the attribution of specimens to a particular genus, or even the naming of the genera. But what is important is the overall pattern. *Dryopithecus* or 'woodland ape', is the stock from which all modern apes have evolved. *Gigantopithecus* was the ancestor of some apparently very large terrestrial apes of Asia that became extinct. The third genus, *Ramapithecus*, is the one in which we are most interested since, as far as one can say at the moment, it is the first representative of the human family – the hominids.

It was *Ramapithecus* that some twelve million years ago stood on the forest fringe poised to face the dangers and the opportunities of the open savanna. Just as the exigencies of tree living forced the pace of early primate evolution, so must the demands of the savanna have speeded the journey from man-like ape to ape-like man. (See Chapter 4 pp. 68ff)

Looking back through the evolutionary sequence of the apes, we see that the most recent specialization has been the hominids, about twelve million years ago; immediately before that the gorilla and chimpanzee diverged from the basic stock; further back are the Asian apes, the orangutan, and the gibbon in that order. The exact timing of all these departures from the basic stock is still controversial, and it has been made even more so by the work of the American biologists, Vincent Sarich and Allan Wilson, who have been examining the similarities and differences between the apes in the molecular constitution of their proteins. Sarich and Wilson do not dispute the *order* in which the apes turned off into their own specialized paths, but they do suggest that it all happened much more recently than the paleontologists believe. For instance, according to the molecular calculations the hominid, chimpanzee, and gorilla lines separated from the ancestral stock four million years ago, the orangutan about seven million, and the gibbon ten million years ago. The confirmation by biochemical techniques of the stages in the evolutionary path is reassuring, but so gross a discrepancy in the apparent timing is more than a little disconcerting. The question still remains to be resolved.

Apart from the apparent conflict over the time-scale of human evolution, studies do emphasize very cogently our relatedness with the chimpanzees. For instance, Wilson, with Marie-Claire King, has compared

The limb bones and skull of Dryopithecus africanus *found in the Miocene deposits of Rusinga Island in Lake Victoria by Louis and Mary Leakey in 1948. It is possible that this species was an ancestor of the chimpanzee.*

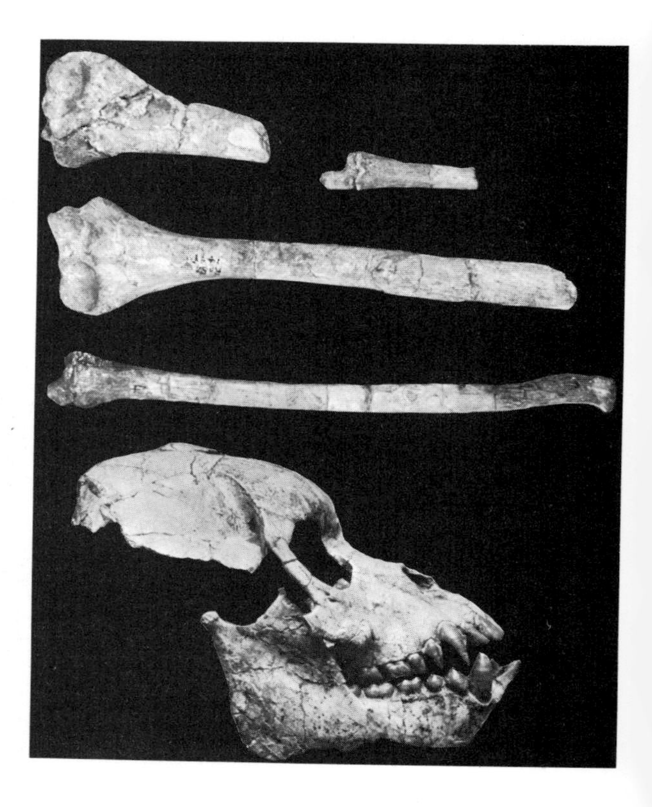

the structure of many proteins in humans and chimpanzees. The difference they find is a small one: 99 per cent of the protein structure is the same! By this test, chimpanzees show closer relatedness to humans than they do to the Asian apes. Even if there were no fossils to back up Darwin's perceptive claim for an African origin of the human race, Wilson and King's results would be powerful support by themselves, as chimpanzees are exclusively African animals.

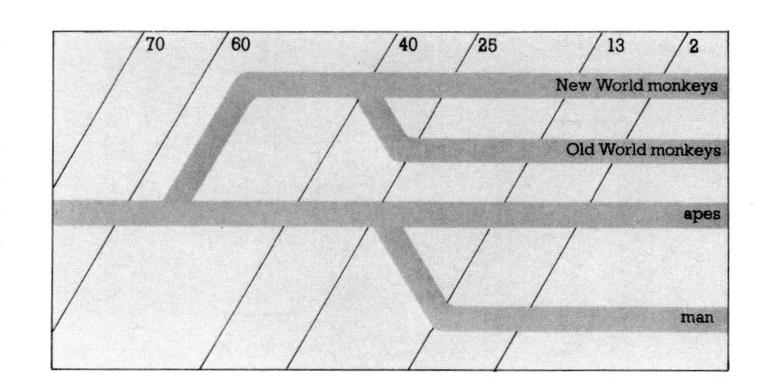

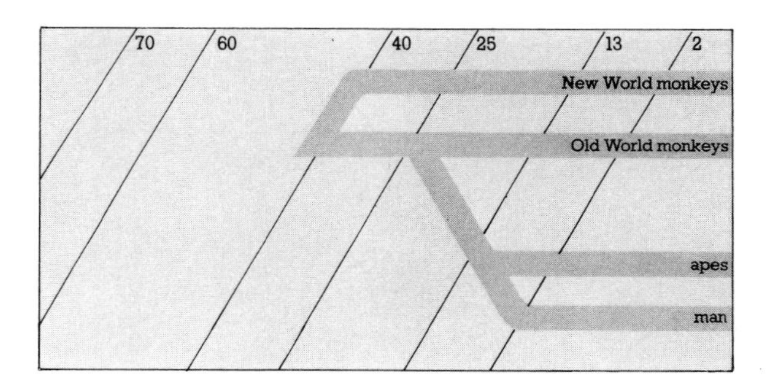

One of the newest clues to mankind's descent is seen in the work of Sarich and Wilson, two American biologists who have been comparing the protein molecules in the blood of various primates. The more alike the structures are, the closer the relationship of the animals concerned. This information has been used to construct the primate evolutionary sequence and on the assumption that the protein has evolved at a constant rate a time scale can be constructed. The diagrams compare their results (right) with the generally accepted evidence (from the fossil record (above right).

A reconstruction of Dryopithecus africanus.

4 Hominid Beginnings

Above all else, we human beings are social animals: emotionally we need to be part of a group, and intellectually we are equipped to understand, and to manipulate, interaction with other people, whether parochially as in personal relations, or politically. In the search for our origins we must be at least as interested in our ancestors' behavior and social organization as in their physical appearance and cognitive capacity. Unfortunately, unlike bones, behavior does not become fossilized, and it is not until after the advent of technology in the evolutionary process that such things as tools and shelters offer the first tantalizing clues to the subject.

All is not hopeless, however, because we can look to our primate cousins for some guidance about the behavior of our earliest ancestors. What factors are important in shaping primate societies, and what kind of adaptations can the animal groups develop? These are the crucial questions with which we are concerned when we try to reconstruct the social behavior of our ancestors. We should look too at the community organization of higher primates, meanwhile stressing its limitations. We are unlikely to find there any real insight into the problem of urban stress, for instance. Nor may we expect to find among nonhuman primates a model of the nuclear family with its tendency toward monogamy. In their different ways both of these behaviors are more likely to be the products of particular economic systems rather than of inbuilt patterns.

One unequivocal message that we do not get from studies of primate societies is that social behavior in each species is to some degree flexible, being fashioned to a great extent by prevailing conditions. That flexibility reaches a peak in the human animal, and we

Previous pages: A group of gorillas in the Congo – the dominant male is on the right. Dian Fossey, who took this photograph, reports that the character of a group is frequently determined by the character of its leader.

Above: The bond between mother and infant chimpanzee is very close. The immature phase is comparatively long in primates and because the juveniles must be cared for they form a stabilizing element in the troop.

Left: Baboons are intensely social animals and each troop has a definite structure. The troop shown here is searching for food away from the trees; the males are more alert for danger than the females and young.

should always bear this in mind when we consider modern societies. Our particular concern at the moment, however, is to arrive at some guidelines for envisioning the sort of social organization that our earliest ancestors might have adopted as they left the forest fringes and began to exploit, and finally to dominate the open grasslands.

The availability of food and probably to a lesser extent, the degree of danger from predators are two basic influences in the relations between individuals within a species; and here the rules that operate for primates also hold for all other species. The higher primates are special because, by their nature, they have a propensity to be social rather than solitary. Differences in environmental conditions therefore tend to determine the *way* in which they are social, rather than *whether* they are social. The reason the higher primates tend to be social animals is that group living offers the opportunity for prolonged learning during childhood; and learning is a pastime in which primates engage far more than any other animal. Learning is a pastime with a purpose, however – namely to equip individuals in a group with a more effective knowledge of the environment in which they must survive. Greater knowledge implies a greater chance of survival – which is what evolution is all about.

A period of childhood education implies dependency on an adult, and this turns out to mean that, without exception, the basis of all primate social groups is the bond between mother and infant. That bond constitutes the social unit out of which all higher orders of society are constructed.

First, let us look at the effects of differences in the availability of food. The two extremes of normal distribution are, on the one hand, food that is scattered, such as roots, seeds and insects; and on the other, localized concentrations, as in a tree loaded with ripe fruit. Food that is dispersed over a wide area tends to be exploited by dispersed social units. By contrast, trees full of fruit will be visited by troops of animals. These are

simple and straightforward rules; no magic formula.

An element of complexity is added, however, when a second set of rules is applied: those concerned with guarding against predators. When animals routinely face the threat of death from predators, they can either organize some form of defense, or flee. Most do both. Defending efficiently usually means living as part of a group. It is the males, who very often are much bigger than the females, and equipped with formidable canine teeth, whose responsibility it is to defend the troop, or at least to hold predators at bay long enough for its members to escape.

Olive baboons give us a neat example of the interplay between dietary habits and the need for defense against predators. From their diet, one would expect baboons to be social dispersers rather than social groupers. But, as they live in more or less open country where the threat of predation is very real, the dietary propensity is to some extent overruled. Baboons feed by squatting to pull up roots and shoots, or to turn over stones in search of insects and grubs, but they frequently look up and glance around, keeping a close watch for signs of trouble. Group living not only allows effective defense in the event of attack, but also gives the means of constant surveillance, such as a solitary animal or a small group could not achieve: in a group of, say, sixty baboons there will always be at least six looking up from their meal at any one time. When predators are particularly numerous, a baboon troop will remain more than normally close to trees where they can quickly seek refuge.

Baboons' habitats vary considerably; some troops live in fully open savanna, and others are to be found in what is best described as open woodland. Although the overall structures of troops in these two different ecological niches are basically the same, social interactions, even in the same species, are markedly different. The principal contrast is in the degree of social tension, which in the savanna troops is greatly sharpened. Troop life among the woodland baboons is flexible, being dominated by a number of mature males who do establish a loose sort of hierarchy.

The pair-bonding shown by the gibbon, far left, is not seen in any other ape. The orangutan, particularly the male, left, tends to be even less social, often occurring as a solitary animal.

There is a great deal of movement of males between neighboring troops, and the integrity of each band is maintained by a core of females and their offspring.

If such a troop were transported to the open savanna, where the danger of predation is greater, it would immediately become much more tightly knit; positions in the hierarchy of male dominance would be keenly contested; transfer between groups would be less common; and the generally heightened tension would lead to a rise in the number of aggressive encounters between individuals. The change in the social interactions in the group would reflect, presumably, the vital necessity for greater vigilance and tighter organization in the event of real danger.

As well as demonstrating the flexibility of social patterns in baboon troops, this example points up very clearly the different demands of particular environments. Chimpanzees have been seen to undergo similar social changes when faced with a switch from secure, relatively closed woodland to more open trees and grassland. Unlike baboons, they do not readily venture into the open savanna. Chimpanzee troops are highly social structures; we are only now beginning to realize, largely through the patient and brilliant observations of Jane Goodall and her colleagues at the Gombe Stream Reserve in Tanzania, how profoundly sociable they are. There is clearly more to chimpanzee life than anyone ever suspected.

In chimpanzee troops, which may include as many as 100 individuals, though the number is usually a bit lower, the bond between mother and infant is clearly the stable element. Males do vie for high status in a hierarchy, and their encounters are somewhat similar to those between competing male woodland baboons. In a chimpanzee troop, unlike one composed of savanna baboons, the dominant male is relatively tolerant of other males' attentions of the females; sexual promiscuity is in the natural order of things. Chimpanzees are basically fruit-eaters, and this means that their troops' density is adapted to the immediate availability of food. When trees are laden with fruit or nuts the whole troop coalesces and has a feast. But often ripe fruits are widely scattered, and for all members to get enough food the troop is forced to fragment into sub-groups, which may be of mothers and children, males and females, or all males.

Periodic dispersion and aggregation of the group to make the best use of available food is an interesting pattern, all the more since it appears in the archeological evidence left by our human ancestors and in contemporary hunter-gatherer communities. For instance, the G/wi people in Botswana coalesce into their bands during the very brief rainy season (literally about eight weeks of the year), to take advantage of the transient standing water that results. For the rest of the year the band is fragmented into nuclear family units, who forage for succulent plants as their only source of water.

Life on the ground is risky. As we see with the baboons, one means of defense is often an increase in the size of the male, a tendency that reaches its extreme in the massive male gorilla, which can weigh up to 400 pounds. On reaching adulthood, gorillas usually have to leave behind all thoughts of climbing trees as they did in their childhood: for most trees they are just too big. But their size means they have no fear from predators – except man, that is. Because their physical defense is so effective, gorilla troops are much smaller than those of baboons, numbering perhaps twenty individuals dominated by an elegant silverback male. Unlike many monkey troops dominated by a single male, in these units the *alpha* gorilla is remarkably tolerant of other mature males, a sign, possibly, of the ape's greater intelligence. For intelligence is frequently as much about *not* doing things as it is about being clever in what one does.

Almost certainly gorillas could survive very well in small dispersed 'nuclear families' if they chose to. Their lack of fear from predators and their diet of leaves would allow them to go their own way. Presumably, however, they do not because, though the group is relatively small, it is still big enough to offer the advantages of social learning.

Because of their rather special ecological adaptations, the Asian apes (gibbons, siamangs, and orangutans) have less to teach us that is directly relevant to the possible social organization of our earliest ancestors. For instance, the gibbon appears to have wedged itself inextricably into an ecological niche. Its spectacular mobility is ideal for plucking fruit from inaccessible sites, but is rather inefficient for traveling over large distances. For this reason the gibbon appears to have been 'forced' to become a territorial animal (unusual among the higher primates), guarding a delineated feeding area against all intruders. The group size is reduced to its minimum of one female, one male, and their offspring. There is no difference in size between the sexes, probably because it is advantageous to have the female acting the role of an extra 'male' in territorial defense.

The gibbon is the only ape which displays so-called pair-bonding, the social pattern many people like to think is natural for humans. Whether or not this is so, the gibbon has arrived at that solution because it is a relatively solitary animal. Humans' ancestors most definitely were social animals. It is probably no accident that, as well as being the most solitary of higher primates, the gibbon appears to be the least well endowed intellectually.

The gibbons' Asian cousin, the orangutan (whose name is Malay, for 'person of the forest') is something of a puzzle. Almost ten times as big as the gibbon, this beautiful golden-haired ape also appears to be something of an antisocial creature, the large males particularly so. A fruit-eater like the gibbon, the orangutan climbs dextrously, but cautiously, through the trees of swampy rain forests in a very restricted area of southeast Asia, Borneo, and Sumatra. The orangutan population has been drastically reduced over the years by irresponsible human predation. and it is possible that the apparent social pattern of small groups (a mother and her offspring, occasionally joined by a male) is an aberration inflicted on a vanishing species. Some people suggest that true orangutan behavior may be much more like the gorilla's and less like the gibbon's than it appears. These questions about the orangutan, the least studied of all the apes, cannot yet be properly answered.

Most of the higher primates respond in biologically predictable ways to important external influences, but such responses are largely inseparable from the basic human propensity towards social groupings. Along with the opportunity that sociability offers for the education of individuals, a collection of individuals also has the potential for developing a 'group wisdom', which is no less vital for the further development of intellectual abilities.

A large male gorilla 'chest-beating' – a display that is almost certainly part of a social mechanism.

The farther up the evolutionary path one travels, the longer the period that offspring depend on an adult. Baboon infants are closely dependent on their mothers for two years, and then enter a juvenile period lasting about twice that long. Chimpanzees' infancy lasts some three years, and they are juveniles for seven more. Compare this with a human infancy of six years and a further fourteen as 'juveniles.' On the other hand, these disparate apprenticeships follow pregnancies of about the same length. This period of prematurity is devoted to learning skills of two sorts: first, the social rules of the group; and second, those skills required for survival as a member of the group, in an outside world that contains both potential food and potential danger, about which information must be acquired. The longer the period of pre-maturity, the more there is that can be learned.

It is safe to say that primates would not be social animals if the trait were not evolutionarily advantageous, and we have identified the advantage as the possibility of learning skills related both to social interactions and to the resources in the environment. One can argue that it is the more efficient exploitation of environmental resources through greater knowledge that formed the thrust of the evolutionary benefit, and that the acquisition of many social skills followed, as a way of enabling the former to be achieved. In other words, for an animal to learn effectively about its environment it must be part of a group, and the group cannot operate efficiently over long periods unless its members can cope with social encounters; hence the need to learn the rules for those encounters. This is a tricky argument, as arguments in evolutionary biology often are, for it is always very difficult to distinguish the prime mover from its inevitable secondary consequences.

Together with learning in a social context, the second major benefit of group living – 'group wisdom' – represents the beginning of culture. As a social entity a troop of chimpanzees, say, has access to a pool of knowledge greater than any one individual would possess. Yet that knowledge is available to each member. The edibility and location of particular foods or the identification of a potentially threatening predator, for example, can become part of group wisdom. And so too can unpleasant experiences. For instance, a biologist who was studying parasitic infections in Kenya once shot two baboons belonging to the same troop from his car. More than eight months later, *all* the members of that troop were still very suspicious of cars, even though only a couple of them could have seen the incident.

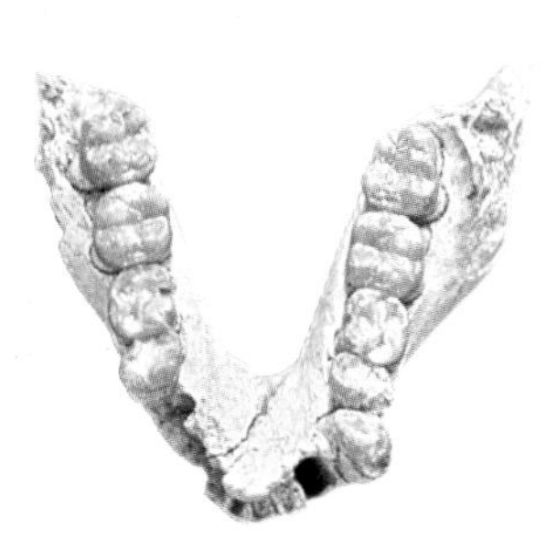

The map, left, indicates the sites where various species ascribed to the genus Ramapithecus *have been found. The lithological nature of the deposits concerned and the associated faunal finds indicate a forest environment, probably having open areas bordering on rivers and lakes. Above left is a partial reconstruction of the lower face of* Ramapithecus *from Fort Ternan and, above right, a mandible found in Turkey. On the right, a reconstruction of* Ramapithecus *which, because so few remains have been found, must be very tentative.*

Another good example of group instruction in baboons comes from the American primatologist Shirley Strum. During her study of a troop nicknamed the Pumphouse Gang, in the Rift Valley 90 miles northeast of Nairobi, she noticed that one of them, a mature male she called Rad, was particularly adept at hunting small antelope, especially Thompson's gazelle. Over a period of two years, most of the males in the Pumphouse Gang had also become enthusiastic about meat-eating. It appears that they had learned by example and had even developed (by chance to begin with) a method for cooperative hunting. What we see here are the rudiments of culture, similar to those manifested by Jane Goodall's chimpanzees who had learned to 'fish' for termites.

What, then, do these observations have to tell us about the behavior of our ancient ancestor *Ramapithecus*? Compared with the fossil remains from Montana of the pioneering primate, the evidence concerning *Ramapithecus* is considerable – though in absolute terms it remains tantalizingly small: fragments of upper and lower jaws, plus a collection of teeth, representing perhaps thirty or more individuals, are all we have from which to piece together a picture of the gradual transition from ape to hominid. We do know, at any rate, that the fossil remains have been discovered as far apart as India, Kenya, Hungary, Pakistan, and Turkey. And we are fairly certain that when *Ramapithecus* flourished, between ten and twelve million years ago, it favored forested areas traversed by rivers, thus providing some open terrain. In the absence of firm evidence, we have to speculate that the rapidly-changing climatic conditions which destroyed the forests and encouraged the spread of savannas must have applied the final pressure for adaptation to open country.

The most dramatic evidence concerning the behavior of *Ramapithecus* so far discovered comes, however, from their canine teeth. In most apes and Old World monkeys these teeth are long and sharp, and this makes them effective for defense against predators and in tearing up large morsels of food, and most important of all in the displays that accompany disputes over hierarchy. The canine teeth of *Ramapithecus*, on the other hand, turn out to be fairly small – not much bigger, in fact, than the neighboring incisors or premolars. What does this mean?

In attempting to find some kind of answer we can first turn for help to a living monkey, the gelada baboon. Limited today to high plateaux in Ethiopia, baboons of this species feed in the open grassland during the day and sleep on rocky ledges among the cliffs at night. They eat an assortment of seeds, grasses, bulbs, rhizomes, and insects which they pick up from the ground – an aspect of feeding behavior that has fostered the evolution of one of the better precision grips among non-human primates. Their teeth have also become adapted to this manner of feeding. In contrast with many monkeys and apes, they have small incisors and large molars. The reason is clear:

During the infant phase, the young primate is very dependent on its mother. In the case of the baboon this period lasts two years and is followed by a juvenile phase of about four years.

since their food does not need to be torn apart, the slicing edge provided by the row of incisors would serve no purpose; whereas they do need to grind it very thoroughly, hence the large molars.

In the gelada baboon, moreover, the canine teeth are relatively small (although bigger in the male than the female).

The reason for the disappearance of these long canines is probably mechanical: they would impede the rotary grinding action that is required for pulping their tough food. It was on the basis of this reasoning, applied to the teeth of *Ramapithecus* that the American biologist, Clifford Jolly, constructed his hypothesis that the original hominids were seed-eaters. Basically he suggests that a shift to an open country, seed-eating diet in the early hominids can explain the important changes in their tooth pattern. Not only would the canines have to be reduced in size to allow efficient chewing, but there would also be a strong selection pressure to develop a precision grip for picking up the small morsels of food. There is undoubtedly some merit in the idea, and the gelada baboon does provide an interesting possible model for this aspect of our evolution.

Were our ancestors of twelve million years ago in fact seed-eaters? From the fossil teeth of *Ramapithecus*, we can infer that whatever their diet may have been, it must have called for a great deal of chewing before the pulp was ready to swallow. Like those of the gelada, the incisors of *Ramapithecus* were diminutive: but they were different in that they were

Shirley Strum, watched by two baboons while taking notes, studied a troop nicknamed the Pumphouse Gang. Top, Rad, carrying a freshly caught Thompson's gazelle and, center, two males at a kill.

Above: Chimpanzees exhibit various displays when excited – they may jump about with hair bristling, brandish sticks or even throw objects.

Left: A baboon displaying its large canines in a threatening attitude. Known as the baboon yawn, it is purely bluff used in social encounters.

scooped into a cutting edge, whose effectiveness was extended somewhat by a slight flattening of the small canines. So it seems clear that even if *Ramapithecus* did eat tough seeds and grasses, its menu also included material that needed to be torn apart before it was chewed.

Some people have argued alternatively that the diet of *Ramapithecus* was based on meat – the slicing incisors being used to tear the flesh and the molars for crushing bones. But the anatomical evidence simply does not support the notion that such incisors could function in a diet that routinely called for crushing tough bones. For the moment, we cannot be positive about what *Ramapithecus* ate. But that is less important than the clear evidence from the organization of the teeth that this first hominid was exploiting a new primate niche: no other member of the order, extant or extinct, has the same dental pattern. It is interesting to compare that pattern with what is known about another primitive ape, *Gigantopithecus*, fossil remains of which have been found in China and India. In this creature – which, as its name suggests, was much larger than *Ramapithecus*, in fact about the size of a modern gorilla – not only were the molars well adapted for crushing tough material, but its canines had also become flattened to give aid in the crushing of food. Presumably its diet was like the gelada baboon's. Knowing as we do that *Ramapithecus* through the course of evolution had lost the large canine teeth of its ancestors, what we cannot and should not do is to look for a *single* explanation. Rather, we must see this development as belonging to a complex of interactions. First, there is the question of defense against predators, one that probably has something to do with the emergence of early hominid characteristics, but which almost certainly has been overemphasized. Occasionally baboons will display their long threatening teeth to a predator. Even a lion may think twice about taking up the challenge. But it is now clear that baboons' large canines are designed more to impress other baboons than to inflict serious injury on predators – they *can* cause nasty wounds, but they are simply too brittle, and they also project at the wrong angle, to be really effective as weapons. On the occasions when male baboons do stand firm and bare their canines in the presence of an advancing predator, they are engaging in a game of bluff. Mostly, though, these ground-living monkeys will make a dash for the safety of the trees rather than go in for a display of heroics. How would our ancestors have fared in the face of large and powerful predators? Unlike *Gigantopithecus*, its small-canined ancient relative, the first hominid was small, just three and a half feet tall, and thus would not have commanded the respect now accorded to the gorillas. As an inhabitant of open woodland, like the modern baboons it would have made the most of the opportunity for escaping to the trees. But it must have been caught out sometimes, and, if our guess about the gradual change in environment is correct, this danger would have increased as the species moved farther out into the open savanna. What then?

For an answer, once again, we can turn to living species for inspiration – this time to the woodland chimpanzee. When members of this species become excited, and particularly as they engage in displays whose aim is to impress other members of the troop, they frequently brandish quite large branches; and there is no doubt that when confronted with a danger from which they cannot escape rapidly, they will use the same technique. As anyone who has visited a zoo will know, both chimpanzees and gorillas display a respectable aim in throwing anything they can find at spectators to whom they have taken a dislike. This is not an aberrant behavior induced by captivity; chimpanzees in their natural habitat have been observed to hurl stones at nearby baboons.

Although these behaviors hardly constitute a serious threat to the health of an approaching predator, the very act of engaging in them produces a certain effect. And their impressiveness is enhanced because the hurling of projectiles and the brandishing of branches are both done from an upright position, which would tend to make the primate appear much bigger than it really is. In a world where impressive displays mean almost as much as aggressive action, *Ramapithecus* might well have developed techniques for protection against predators similar to those of chimpanzees and gorillas such as would demand an upright posture.

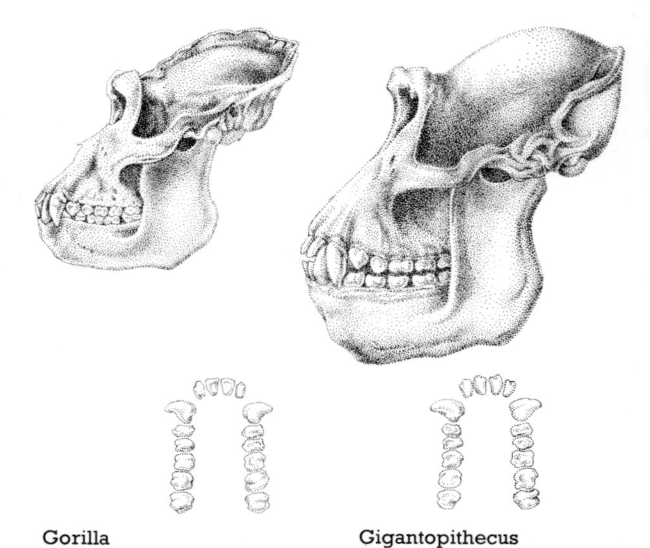

Gorilla **Gigantopithecus**

Surveillance is another aspect of anti-predator behavior that would tend to favor a standing posture. For animals living surrounded by long grass, the ability to stand up and thereby achieve a wider view of the surroundings is a notable advantage. Among baboon troops it is a normal part of daytime surveillance for individuals periodically to balance briefly on their hind legs so as to gaze around looking for potential hazards. An easily maintained upright posture has enormous survival advantages, though by itself surveillance would have been inadequate to explain the evolution of full-time upright walking.

A third factor that may have fed into the 'upright walking equation' is the increased ability to carry objects such as food when the hands are freed. Chimpanzees often move with their arms full of ripe fruit to a suitable place for a meal, though the distance they can waddle – for theirs is not a fully accomplished stride – is only a few yards. One of the burdens a female *Ramapithecus* would have had to carry, of course, is her baby, and this presents a salutary illustration of the feedback loop involved in the early process of hominization.

As the hominids gradually became more bipedal, so the feet would have become adapted for efficient walking. But a foot that can walk is not one that can grip, and this applies to infants as well. As a result, the

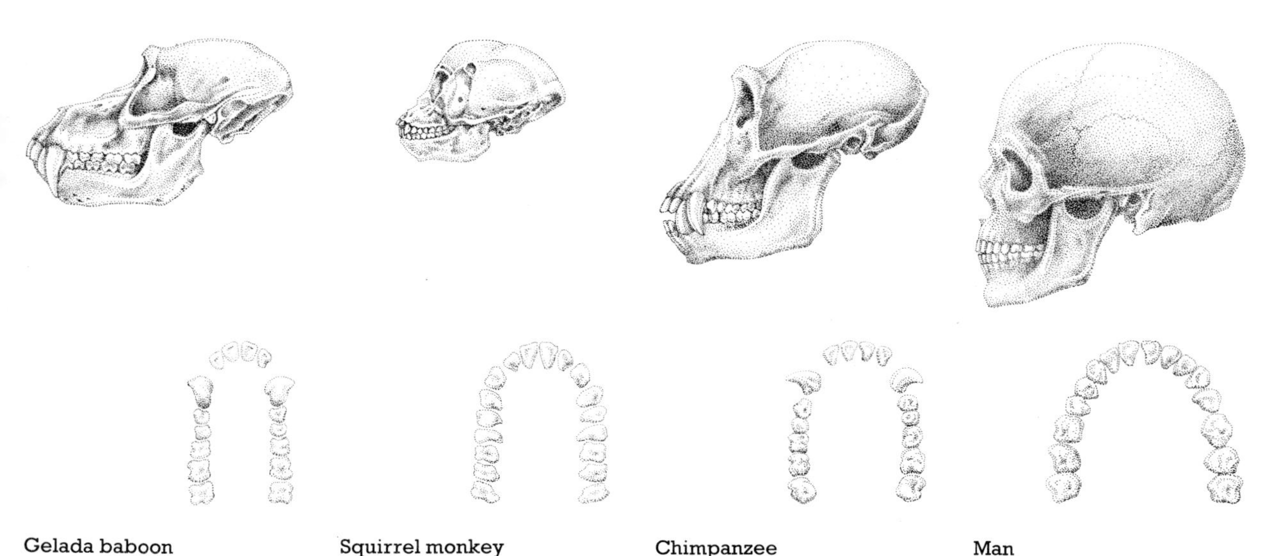

Much can be learned from teeth: their anatomy reflects, among other things, dietary preferences. From our knowledge of the form and function of teeth in living primates we can discover much about the likely food and habits of our ancestors. The illustrations above compare the incisors and molars of the gorilla, Gigantopithecus, *the gelada baboon, the squirrel monkey, the chimpanzee and man; the jaws are not to scale.*

Below: many characteristics have evolved as a result of the positive feedback mechanism. The diagram below shows this in a simplified way. Basically, it is the way in which one character is reinforced by another. For example: bipedalism led to the freeing of the hands – for carrying and for manipulating objects – and the more the hands were used the more efficient upright walking had to become.

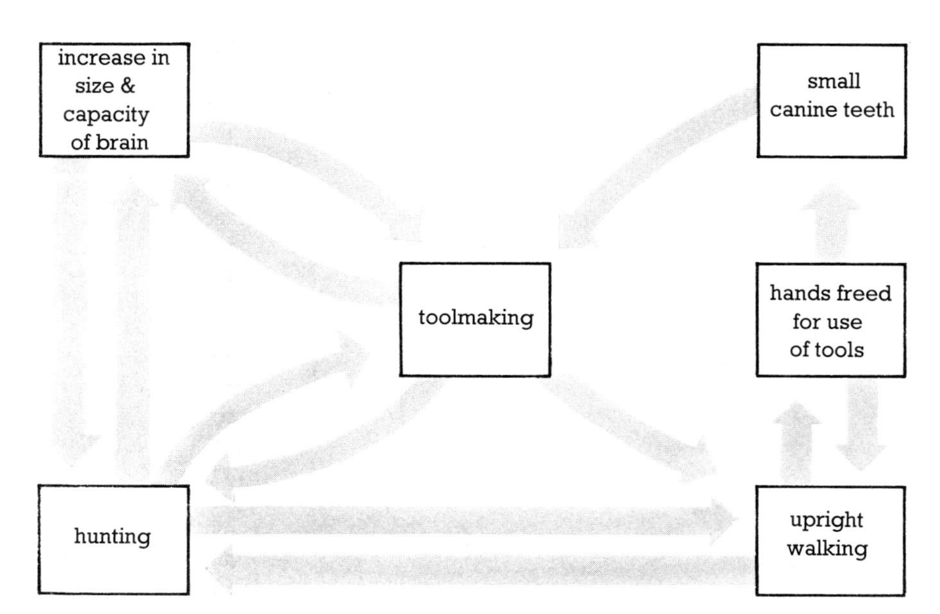

The battered cobble found by Louis Leakey at Fort Ternan among the fossil remains of Ramapithecus.

infant can no longer hold onto its mother's body with two hands and two feet in the same efficient way as its ape cousins. Consequently, the mother has a greater need to be able to carry the infant in her arms, and so efficient upright walking becomes still more important. In other words, the more a species evolves bipedal walking, the more it is forced to evolve still further in that direction. Positive feedback loops have been seen to operate in the development of other activities, but none so neatly as this one.

A favorite suggestion about the object with which an upright early hominid might have filled its now empty hand is, of course, that it might be a tool. Sherwood Washburn, an American primatologist, goes so far as to suggest that the use of tools was the major force in causing the hominids that were our ancestors to take to their hind legs and walk. The idea is certainly attractive, for a number of reasons. As they explored their new ecological habitat, the early hominids would have found the extra behavioral dimension – for that is what tool-using is – a distinct advantage. As well as brandishing sticks or throwing stones in self-defense, the earliest hominids may have discovered the advantage of using tools in procuring food. For instance, they may have broken bones open to get at the nutritious marrow inside, or used a stick to dig up roots. It would certainly have been surprising if *Ramapithecus* did not discover the benefits of using a stick to extend its effective reach.

The only evidence for any of this – and it is exceedingly tenuous evidence – is a discovery made by Louis Leakey among the fossil remains of *Ramapithecus* at Fort Ternan. It consists of what was *possibly* a battered cobble, lying next to an animal bone that *may* have been split by it. If so, the inference would be that animal food was included in the diet of hominids. Thus we don't know whether bone-smashing was routine for *Ramapithecus*, or the isolated exploit of a particularly venturesome individual, or whether the discovery means nothing of either sort, but is no more than an accident of nature.

If *Ramapithecus* did use tools, the greater part of them are much more likely to have been made of wood rather than stone, at least initially. Branches that had been deployed in protective displays could very easily be redeployed in the search for food. The two activities are likely to have developed together, with protection probably leading the way. If wooden implements really did play an important role in the emerging culture of *Ramapithecus*, our problem becomes more difficult in that wood is only rarely preserved and fossilized. Usually it vanishes, leaving us to search for nonexistent clues.

It may well be, however, that the importance of tools in human evolution has been greatly exaggerated, since for so long a time stone tools offered practically the only available glimpse into the lives of our prehistoric ancestors.The image of man the tool-maker has therefore been impressed deeply on the minds of many people. Indeed, Darwin himself wrote in *The Descent of Man*: 'The early male forefathers of man were probably furnished with great canine teeth, but as they gradually acquired the habit of using stones, clubs, and other weapons for fighting their enemies or rivals they would use their jaws and teeth less and less. In this case the jaws, together with the teeth, would become reduced in size.' This scenario clearly combined upright walking and tool-using as the triggering forces of human evolution, with the small canines following as a consequence – the reverse of what now appears likely to have been true. But Darwin's perspicacity in placing the reduced size of the canines at a very early stage in evolution is still nothing short of remarkable, for the first example of our ancestor's diminutive teeth would not be unearthed until almost a century later.

Turning once again to the baboons for inspiration about our distant ancestors' possible behavior, we recall that when they move from woodland to savanna their social grouping becomes tenser and more tightly

Right: A male chimpanzee begging for food from another who is holding a bushbuck.

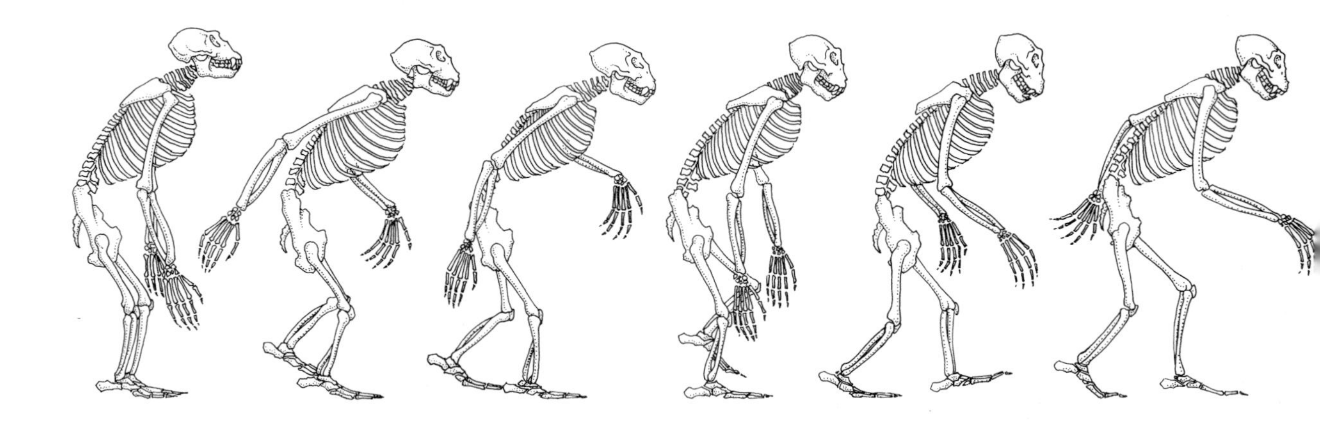

Bipedalism is found much more in apes than in monkeys. Upright walking in the chimpanzee for example, is far less efficient than in man Man's greater capability in this respect is the combination of a large number of anatomical adaptations.

knit. We can speculate that the same kind of tightening would have occurred in troops of *Ramapithecus*. More than ever the defenders of the troop (almost certainly the males) would be welded into occasional cooperative alliances as a way of strengthening their defenses. Collectively some display or threat of force might have been extremely effective, whereas a single branch-waving *Ramapithecus* would probably not be particularly impressive. And at least part of the key to successful defense is simply to be impressive. As with baboons and other primates potentially at risk from predation, we may suppose *Ramapithecus* males to have been bulkier than the females.

The possibility of rising tension among hominids with small canines venturing into relatively open territory would therefore be extreme. Social tension can be valuable in helping to sharpen awareness of interactions between individuals. But it is potentially disruptive if it becomes too great. Moreover, those groups of hominids who did not adapt to coping with the greater potential lethality of their tool technology might well have wiped each other out in an orgy of social conflict. Is it possible, therefore, that the greater than normal cooperation required for defense might be just one method through which this potential tension could be dissipated? Could group cooperation possibly have emerged as a trait in this small-toothed social primate as it ventured into a new, potentially hazardous ecological niche? We believe it could. Indeed, there is a case for recognizing social cooperation as the key factor in the successful evolution of *Homo sapiens*.

There is, of course, one very special activity that is a particularly potent stimulus for group cooperation, and that is the sharing of food. For very simple reasons, animals that are vegetarians do not share their spoils. Fruit and leaves come in small packages, and an individual belonging even to a very cohesive group will simply pick and eat its own fill – that and not a bit more. When a meal is highly concentrated, as it would be in the form of the dead carcass of an animal, the opportunity – and indeed necessity – for sharing arises. For instance, that exemplary social carnivore, the African wild dog, shares its catch among members of its 'camp' who have not taken part in the hunt. It is especially intriguing to discover that chimpanzees, in their occasional forays into meat-eating, share out morsels to individuals who beg persistently enough. This contrasts dramatically with the chimpanzees' more usual vegetarian behavior during which they share their food in a limited way.

Hunting among non-human primates is decidedly unusual, but we now know that it is practiced much more than was once believed. For instance, baboons also have the occasional feast on hares or young ante-

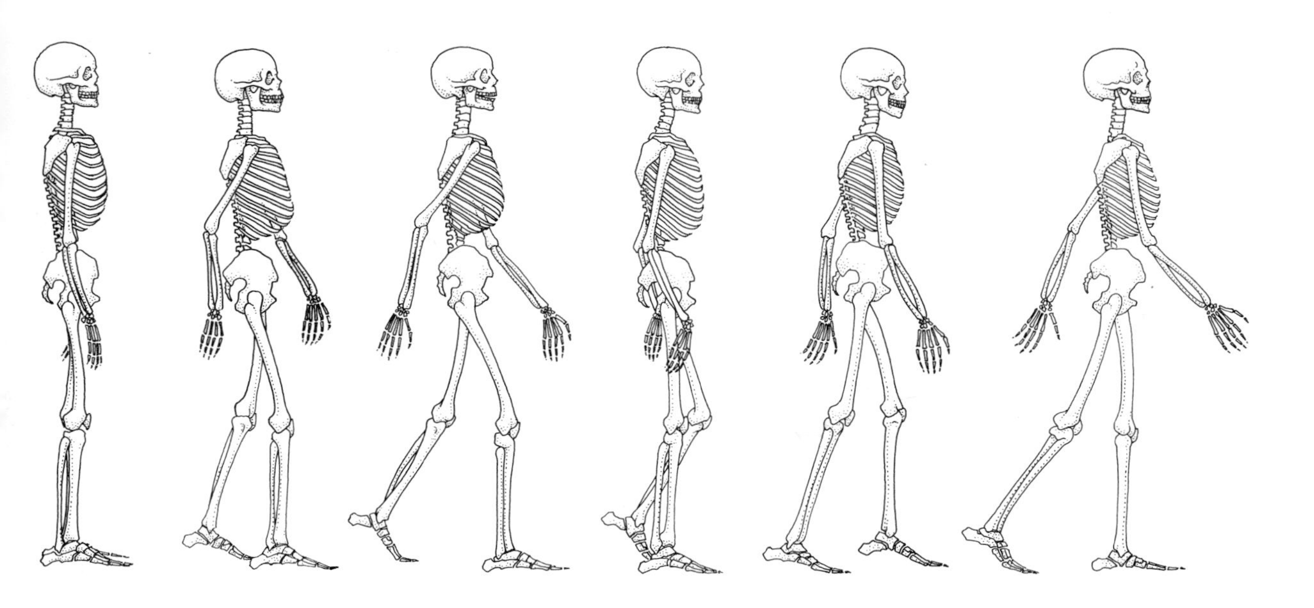

lope. But unlike chimpanzees they do not appear to have developed the sharing habit. Did early *Ramapithecus* eat meat? Very possibly these animals were more nearly omnivorous than their other ape cousins; but it seems likely that, for a start at least, their excursions into meat-eating would have been infrequent rather than routine. What we do know is that some of the descendants of *Ramapithecus* took a great interest in meat, so that by around two or three million years ago animals formed an important part of some hominids' diet. Meat-eating led to sharing, which in turn encouraged greater social cohesion. Precisely when that cohesion began to be important, even in its simplest form, for the moment remains a mystery.

Once some traces of the limbs of a fifteen-million-year-old *Ramapithecus* have been unearthed, as they must surely be soon, we will at last have a real glimpse into the origins of hominid behavior. When did our ancestors stand up? That is still the crucial question.

Because of its mechanics, upright walking is less efficient than four-legged locomotion. We therefore have to suppose that the total behavioral strategy of our emerging ancestors was advantageous enough to overcome this handicap. *How* upright walking became a favored mode of getting about, however, is a much easier question to answer than *why* it happened. Basic arboreal arm-swinging is probably the most favorable pre-walking way of moving: long arms, a short rigid trunk and feet capable of standing are qualities needed for both activities. The transition would therefore not have been physically traumatic in evolutionary terms: a little shortening of the arms, a straightening of the foot, and a broadening of the pelvis would do the trick. Why it happened, though, must remain guesswork – for the moment at least. We can speculate that factors such as the small canine teeth, changes in social organization implied by longer infant dependency, the need for unusual modes of protection, and the use of tools, are all involved in the emergence of upright walking. No single one of these could account for it alone.

If it seems remarkable that evidence concerning the earliest hominids is so sparse, still more remarkable is that there exists virtually no trace of their descendants, over a period lasting from about ten million years to five million years ago. And after that five-million-year gap, the first glimpse consists of a single jaw fragment of uncertain provenance. Not until the period beginning three million years ago does any real, solid evidence appear. By then, it seems, there was not just one hominid type, but several. To differing degrees, these probable descendants of *Ramapithecus* had mastered the art of upright walking; some at least were producing stone tools.

The more than welcome profusion of fossil discoveries during recent years has transformed our view of human evolution. Its course is clearly much less direct and straighforward than was once thought. But as we shall see in the next chapter, it is much more logical too.

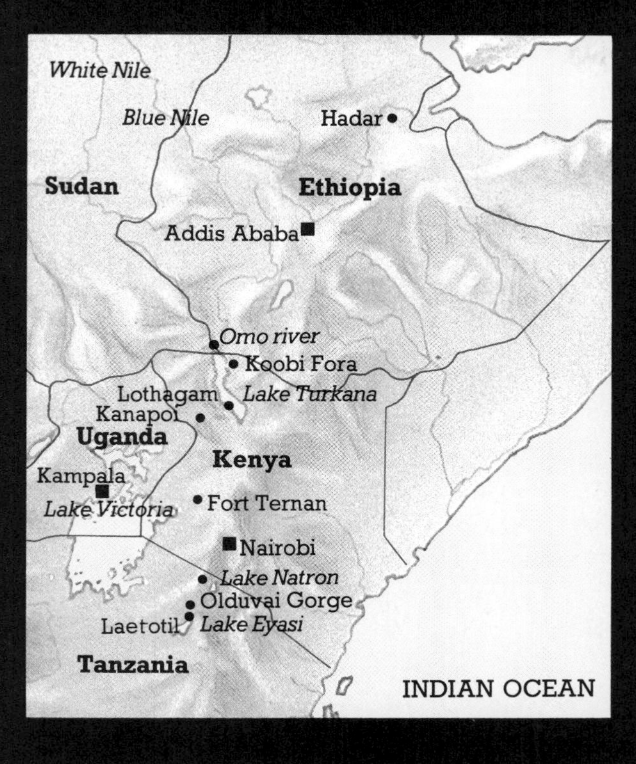

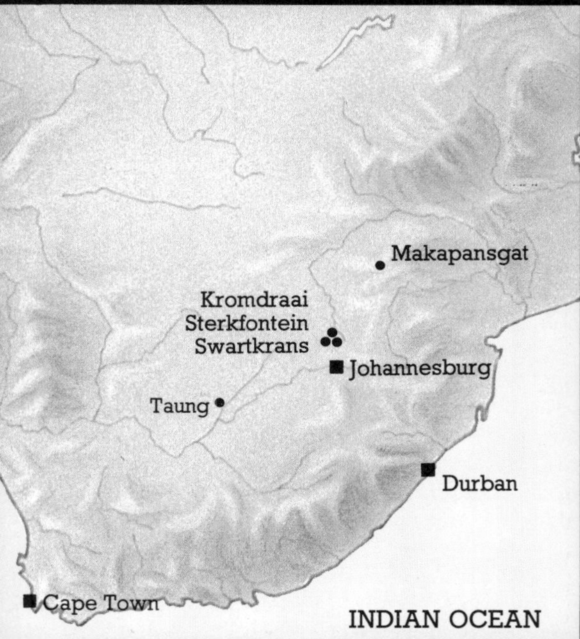

5 The Cradle of Mankind

One late autumn afternoon a band of hominids was moving along the shore of Lake Turkana in northern Kenya. They were looking for a suitable place to camp. A spot was found that was both conveniently close to the cool glistening waters of the jade-green lake and was free of the relentlessly sharp spike grass that makes much of the shore a less than comfortable place on which to rest. The camp was in fact on the sandy bed of a dried-up stream, part of the delta of a small sand river that in the rainy season carries gushing torrents into a main stream, and then into the lake. For four days the small group enjoyed the comfort of the smooth sandy bed, they fed on fish and crocodiles that they caught in the nearby waters, and they feasted on chunks of meat scavenged from a zebra that had been killed and abandoned by lions. At the end of the four days the band moved on, probably to set up another temporary base further along the shore.

The rains came and the rapidly-flowing waters soon covered the remains of the abandoned site with fine silt, and the slow process of burial began, preserving the fragmentary clues that were all that was left of what once had been a busy and organized camp. The scene was now set for someone to come along later to excavate the site and interpret the buried clues. And that is just what happened.

The excavator, Diane Gifford, began carefully uncovering the campsite in the summer of 1974, less than one year after its occupants had moved on. The hominids were modern humans, members of the Dassanetch tribe, a Cushitic pastoralist people who inhabit the north-eastern shores of Lake Turkana, and spread up into Ethiopia. Why excavate modern living sites? Because, as is becoming increasingly obvious, the knowledge of preservation and fossilization processes we acquire from doing this, helps us interpret and understand better the scattered and meager evidence that turns up on ancient campsites. This seemingly odd mixture of ancient and modern is a key part of the research approach that is emerging from the team study of our ancestors who thrived at least two million years ago in areas where the Dassanetch now live.

The link between then and now is made particularly pertinent by the startling similarity in setting between Gifford's Dassanetch camp and an ancient living site located just a few kilometers further north along the shore. There, at what more than two million years ago was the lake margin, a small band of hominids made a temporary home base, just like the Dassanetch group. And the geological evidence tells us that, again like the Dassanetch, this group of people chose a dried stream bed in which to make their camp, a site we now call KBS, after its discoverer Kay Behrensmeyer. Were they, too, escaping the discomforts of spike grass? Possibly. We do know, however, that in the environs of their camp the hominids would have had the welcome possibility of leafy shade in which to escape the burning glare of the sun. And, adding a nice touch of poignancy, one of the leaves that fell into the camp, or perhaps was taken in purposely, left a cast of its shape – it was a fig leaf!

Previous pages: The sites in Africa where the remains of early men have been discovered.

Leaving aside the more fanciful notions of what this campsite with its fig leaf might tell us about the legend of Adam and Eve, we can be sure of one thing: the discoveries on this site, and the hundreds of finds in neighboring sites along the eastern shores of Lake Turkana, combine with the exciting new fossils unearthed by other research teams further north in Ethiopia and further south in Tanzania to transform our view of the path of human evolution. The rich vein of hominid fossils that is currently being tapped in Africa is giving us just the sort of information we need to be able to draw a picture of the evolutionary history of our ancestors, of ourselves, and it is a picture that is now more fact than fantasy. Not very long ago there were almost as many fossil hunters as there were hominid remains found. Now, although there is not exactly an embarrassment of fossil riches, there is certainly an impressive and rapidly-growing array of fossil material for prehistorians to pore over.

The simple, but powerful, message that collectively the recent African discoveries have for us is this: the hominids who, two and a half million years ago, were eventually to evolve into modern humans shared their world with close cousins, hominids that for one reason or another were destined for evolutionary oblivion. At one time popular notion held that the various primitive pre-humans all fitted into a simple scheme of steady progression from an ape-like stock right through to modern humans. This notion thrived because it buttressed the view, however unconsciously it was held, that the ultimate emergence of *Homo sapiens sapiens*, of us, was somehow predetermined. All the 'near-men', 'pre-humans', 'ape-men', 'men-apes', or whatever they were severally called, merely represented steps along a path, and at the end of the path there was the ultimate perfect product. All very understandable, but not very sound biologically.

As we see it, the story of human evolution some two to three million years ago involved probably four main characters. First in our list, but not necessarily first in importance at the time, was a hominid whose principal distinction was that its descendants are us, modern humans. Although no one can be sure, because there are no complete skeletons, this ancient forerunner of true humans probably stood around five feet tall; moreover it stood fully erect, just as we do today. The size of its brain was around two thirds of the average human's today. Looking not dissimilar, but more diminutive and with a smaller brain, was the second hominid in our list. Third was a much stockier individual, standing perhaps five feet tall, but with a significantly smaller brain than our direct ancestor's. Last is the smallest creature of them all, and the one most shadowy in terms of the fossil record (except perhaps for one remarkably-complete skeleton discovered recently in Ethiopia).

These, then, are the actors in the story. What are their names? In answering this we run smack into one of the most thorny problems that has bedevilled human prehistory ever since the science got underway. During past years fossil finders frequently attached labels to their discoveries on the flimsiest of anatomical nuances. New species and genera were created with little regard to the inevitable variations between individuals of the same species, and the possible biological relationship between one species and another. For a group of individuals to constitute a separate species they must be biologically distinct from their relatives, and this may include subtle but important anatomical adaptations to particular environments. Certainly, in the wild, interbreeding between species is very rare.

Now, as we are dealing with creatures long dead, the opportunity to test the separate species hypothesis through information on fertility is denied us. Anatomy is our only guide, and anatomy of fossilized, usually fragmented, bones at that. No wonder, then, that it was as easy to set up a new species as it was to challenge someone else's claim. The only way out is sophisticated statistical analysis of bone measurements and form, so that slight differences between specimens may be classified as either falling within the range of normal variation, or as being biologically significant.

Bearing these caveats in mind, we can begin to put some names on our list, starting with the middle two for the best reason of all: they are easiest. Because of important similarities between them, they are both lumped into the same genus, *Australopithecus*, but the gracile individual is given the specific name of *africanus* while for the robust creature we shall stick to the specific name of *boisei*. The labels we use are at worst convenient, but without being misleading; some people argue for a different naming scheme, but that is not especially important here.

The first character in the list is more of a problem. Although anatomically not *startlingly* different from its hominid cousins there are good reasons for believing that there was a significant behavioral gulf between them. And it was behavioral sophistication that in large measure helped to propel this human ancestor along the road to mankind. In evolutionary terms the speed at which this journey was undertaken was breathtaking: biological milestones were reached and passed with great rapidity. The creature was in such a dynamic state of evolution that it is almost more trouble than it is worth to try to pin down any particular stage by attaching definite specific names. The major point, however, is that this creature was on its way to modern humans, *Homo sapiens sapiens*. For the moment we will apply the generic name of *Homo* without tying ourselves down with specific names. This may seem to be a pedantic point, but, as will become clearer as we explore the bones and stones that litter the path of human evolution, it is important for a rational view of how that path was traveled.

Last in our list is a link with the past, the most ancient human ancestor of all, *Ramapithecus*. We left *Ramapithecus*, or rather the meager collection of fossil fragments that we know him by, tentatively exploring the forest fringes some nine to twelve million years ago. There then opens up an enormous fossil void until round about four million years ago. And it is not until the two- to three-million year stage that there is anything like enough hominid fossils for anyone to have a sensible conversation about. This yawning void is particularly frustrating because on one side of it there is just one creature, *Ramapithecus*, while milling about on the other side is a menagerie of hominids, *Australopithecus africanus, Australopithecus boisei*, early *Homo*, and late *Ramapithecus*. And if we were to leap forward in time to perhaps three-quarters of a million

Above: Three Dassanetch men at a temporary fishing camp on the beach at Lake Turkana. Diane Gifford has been excavating abandoned campsites of the Dassanetch tribe. The results of her studies help us to interpret and understand the ancient campsites that are found by archaeologists. The campsite, above left, was abandoned in November 1973 – the day before it was photographed. Field Assistant Andrew Kilzono (right) points to hearthstones and a woodpile. Assistant Jack Kilzono holds the remains of a large crocodile. The campsite was examined again after the first rain in April 1974 and the photograph, far left, shows Andrew Kilzono pointing to the hearthstones that are now projecting from the newly deposited silt. The crocodile remains lie behind the left end of the meter stick. In August 1974 the campsite was excavated and the photograph, left, shows Kay Behrensmeyer looking at the stratigraphic section while Andrew Kilzono prepares to excavate the block of deposits containing the hearth. The remains of the woodpile are in front of Behrensmeyer and those of the crocodile are behind and to the left of Kilzono.

years ago, we find that the menagerie has dwindled once again to a single representative, a creature called *Homo erectus*. The story of that initial diversification followed by a drastic pruning is the story of human evolution.

It is interesting to try to piece together something of the plot of this story and make some guesses about the habits and interactions of the main characters. But there is a twist in the plot, and it is a curious geographical one. We know that *Ramapithecus* was not at all a parochial creature: he lived in what are now Europe, Africa and Asia. We know, too, that *Homo erectus* occupied these same geographical areas. But the key elements of the story, the diversification and pruning, the period when different hominid types lived side by side, appear to have been played out only in Africa. It is not impossible that fossilized remains of australopithecines lie buried in the hillsides of Europe and Asia, but it is more than a little surprising that so far no one has found any convincing evidence of them (there are claims of fragments of *Australopithecus* from Java and China, but the specimens are so poor as to defy confident interpretation)

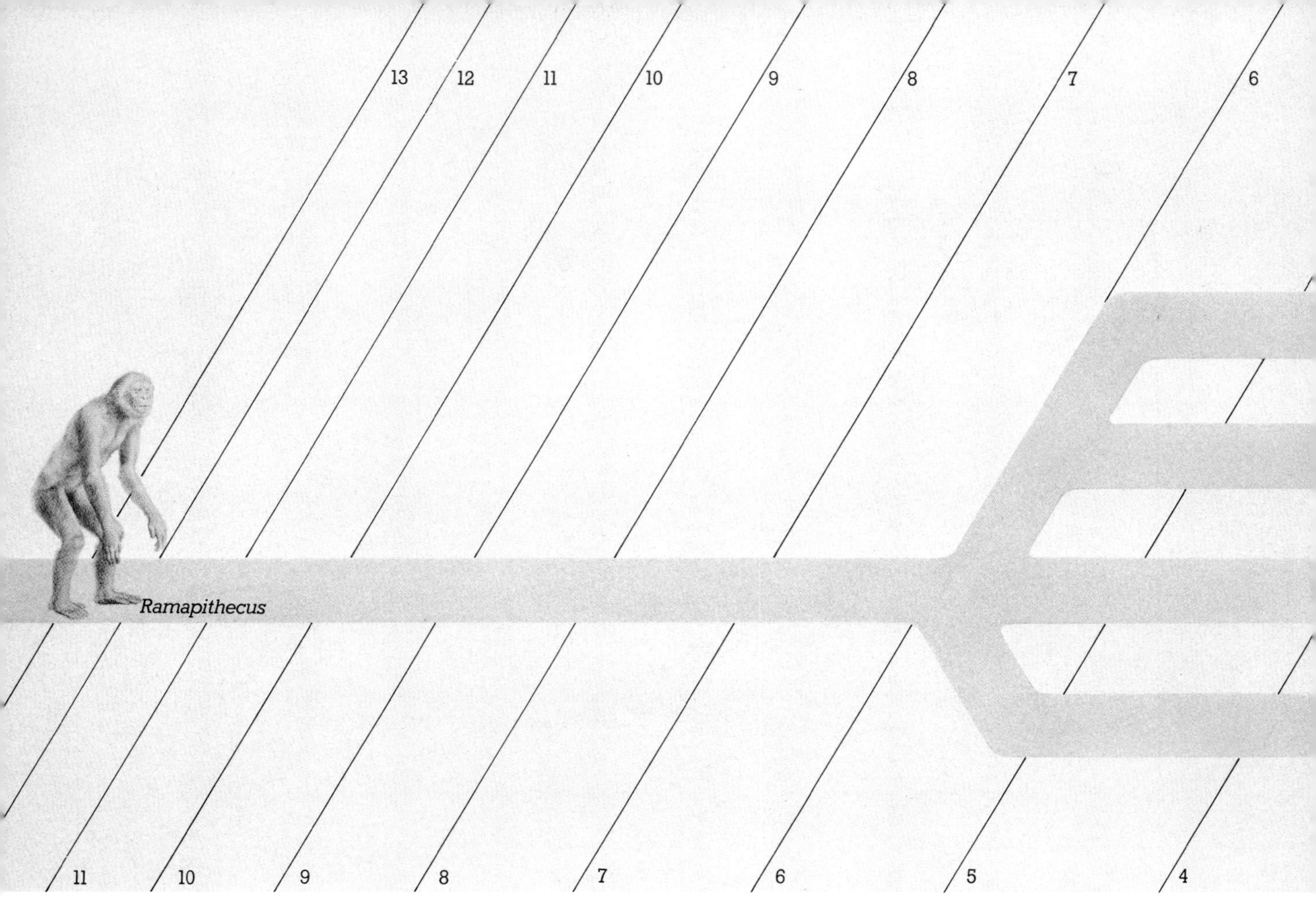

Why was it that the basic ancestral stock, *Ramapithecus*, diversified to generate the members of the hominid family? And why only in Africa? Why was the *Homo* line so spectacularly successful? And what forces nudged the two *Australopithecus* species into extinction? These are the tantalizing question marks that hang over our prehistory. The urge to know what happened is very great, an irresistible inbuilt curiosity about our origins. What was it actually like to be almost human, to be making tools of wood and stone, to live in organized social groups, and to share the world with creatures who were also more human than ape, but who lived quite different lives? If we are honest we have to face the fact that we shall never truly know. We can guess, and the broad research approach encompassing study of the ancient and the modern is beginning to give those guesses some substance. But even if our guesses were absolutely right, there is no one to say, yes you have the correct answer! Inescapably, it is a matter of faith, and this makes the whole problem more challenging – and more exciting.

Our task is not unlike attempting to assemble a three-dimensional jigsaw puzzle in which most of the pieces are missing, and those few bits that are to hand are broken! The jigsaw is multidimensional because, against a background of the physical evolution of our ancestors, we are also trying to construct some semblance of their social and behavioral patterns.

The core of the problem, then, is the fossil record, the fragments of bones which survived the combined destructive activities of the environment to become preserved for later discovery by one of the many teams of fossil hunters now scouring promising sites in Africa. Just as finding a key piece in a jigsaw can help you slot into place many others that have been something of a puzzle, so the lucky discovery at the East Turkana site of a remarkable fossil at the end of 1972 proved to be an important event in the emergence of the new picture of human evolution. In many ways the fossil skull, which is usually referred to simply as 1470 after its index number at the National Museums of Kenya, merely confirmed certain earlier ideas of the

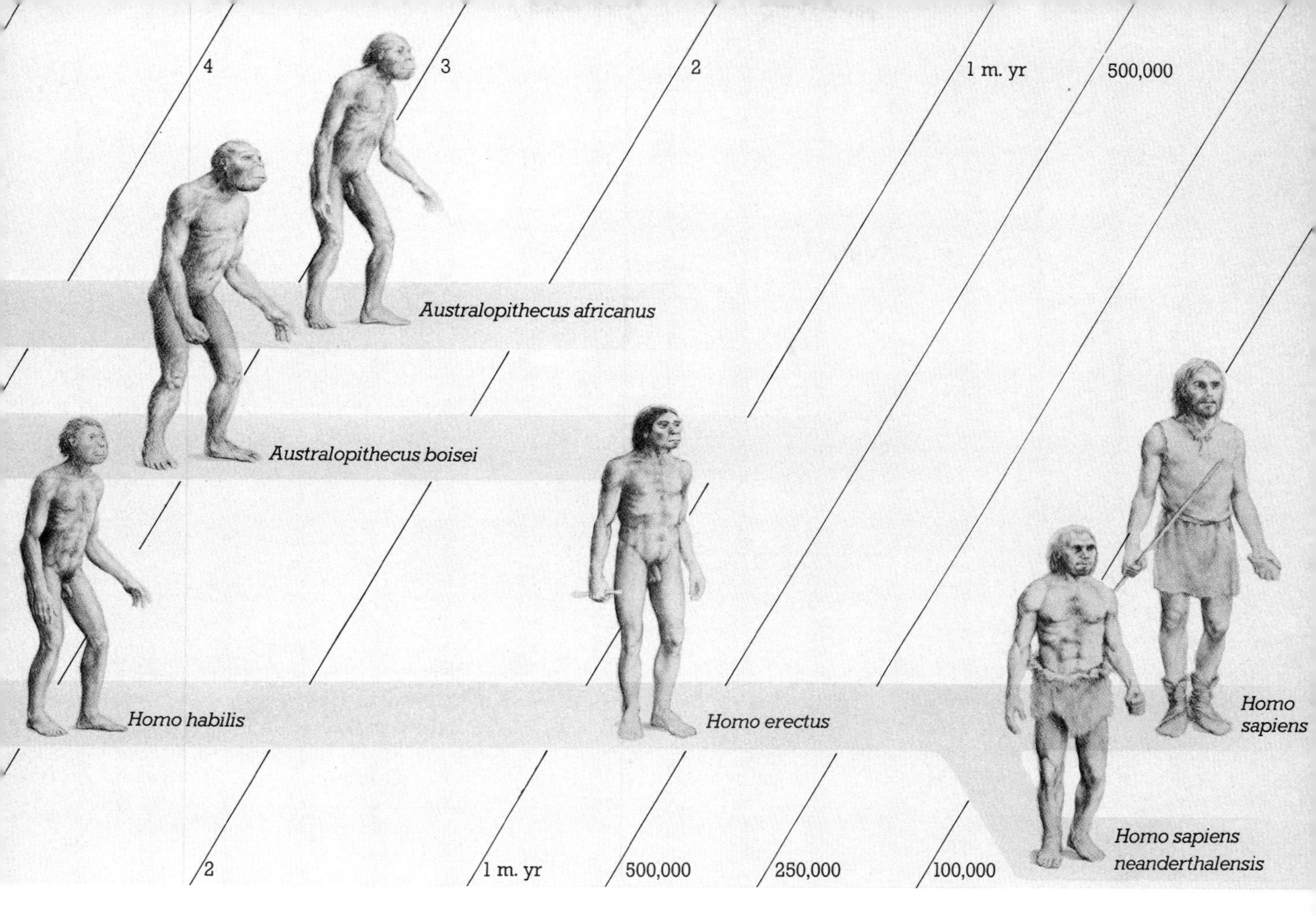

The probable evolutionary path of the hominids. In the earlier days of the theory of evolution as applied to man, the various discoveries were placed on a straight path without any branches. Since then several schemes have been put forward, and although there has never been total agreement on the exact nature of the path – perhaps there never will be – the picture we build becomes more and more accurate as knowledge is gained through more discoveries and the use of more scientific disciplines.

path of evolution, ideas that seemed reasonable in terms of a biologically-sound scenario of human prehistory, but ones that lacked the important evidence of a reasonably complete specimen.

In terms of fossil discovery 1470 has a predecessor, an individual who was found early in 1961 in Tanzania's famous Olduvai Gorge. That find was important because, although the brain case was not complete, it was obvious from its size and shape that here was a very advanced hominid who had lived around one and three-quarter million years ago: he was eventually called *Homo habilis*. This was the first evidence that early members of the human lineage were contemporaries of the australopithecines, not descendants as was generally believed. Exciting though the discovery of *Homo habilis* was, it remained frustratingly incomplete. For the newly developing theory of human evolution to be really persuasive there needed to be a better, more complete specimen discovered. That turned out to be 1470.

Like the Olduvai *Homo Habilis*, 1470 has a large

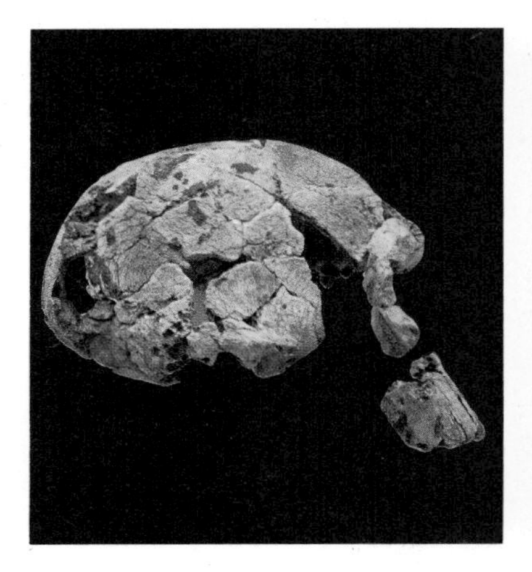
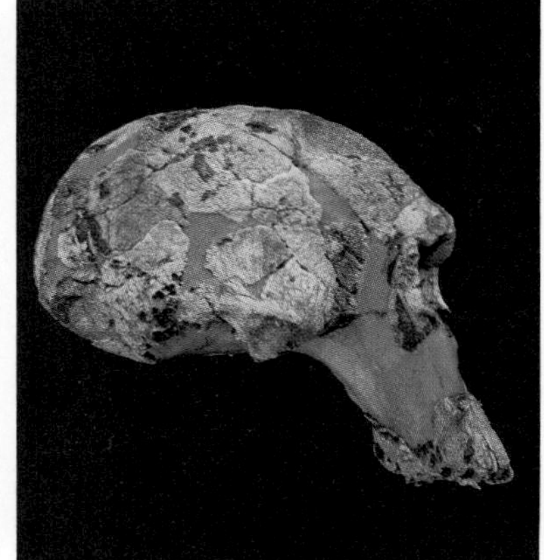
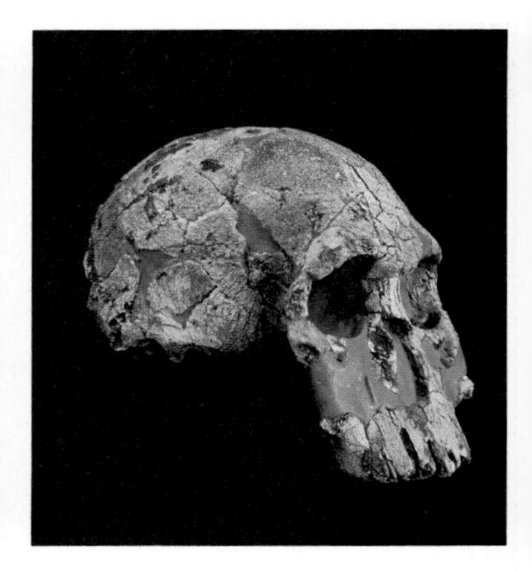

The oldest 'complete' skull – 1470 – found on the eastern shore of Lake Turkana. On the left is a side view of the first construction, and alongside the same view of the preliminary full reconstruction with the gaps filled with plasticine. Above is a three-quarter view from the front and, right, a full frontal view. The skull is very suggestive of a highly evolved hominid: the principal feature is its relatively large brain, about 800cc (more than half a modern human's). 1470 is close to two and a half million years old.

skull, and it too can unmistakably be placed on the path to modern humans. Indeed, there is every reason to classify it as a *Homo habilis*: the Olduvai and Turkana skulls are remains of the same species. But the fascinating point about 1470 is that he lived at least two million years ago, and possibly nearer to three. And he had a bigger brain than the original Olduvai *Homo habilis*.

Unlike the first hominid skull to be found in the shoreside deposits of Lake Turkana in 1969, which was exposed virtually intact in a dried stream bed just waiting to be picked up, 1470 came out of the ground in scores of tiny fragments. Bernard Ngeneo, a member of the fossil-hunting team, had spotted just a few scraps of bone being eroded out of sandy sediments in a steep gully in an area that now looks more like a moonscape than a place that once supported a vital stage in our ancestral history, and eventually the rest of the shattered skull was sifted out of the sediments, giving Meave Leakey and British anatomist Bernard Wood the daunting task of putting it all back together.

Even before the reconstruction was complete, a task which took six weeks of patience and dexterity, excitement was running high. The reason was that two of the larger fragments from the crushed cranium – two pieces from the frontal region – were enough to indicate that this was something special, an individual that displayed signs of evolutionary sophistication which, according to the thinking of the day, was just too advanced for its age. After those six weeks, when enough of the pieces of bone had been slotted into place to give us a reliable idea of what 1470 looked like, it was clear that those two frontal fragments had not held an empty promise: 1470 *was* very special – it represented an almost complete skull of *Homo habilis*.

This remarkable skull confirmed two things. First, that the human ancestral line, *Homo*, originated much earlier than most people suspected, earlier perhaps by as much as a million years. Second, because the history of *Homo* goes back that far, it means that these individuals were living at the same time as some of the earliest australopithecines, making it unlikely that our direct ancestors are evolutionary descendants of the australopithecines – cousins, yes, but descendants, no. Up to that time workers in this field believed that although the robust *Australopithecus* might be a sidetrack of the main path of human evolution, its slighter cousin, *Australopithecus africanus*, was certainly marching along the main route, eventually to give rise to the *Homo* line.

With this crucial piece of evidence now in our grasp, the newly-emerging theory of man's origins gradually gained strength. It was now possible to predict that some day fossil hunters would unearth early specimens of *Homo* individuals perhaps as old as four or

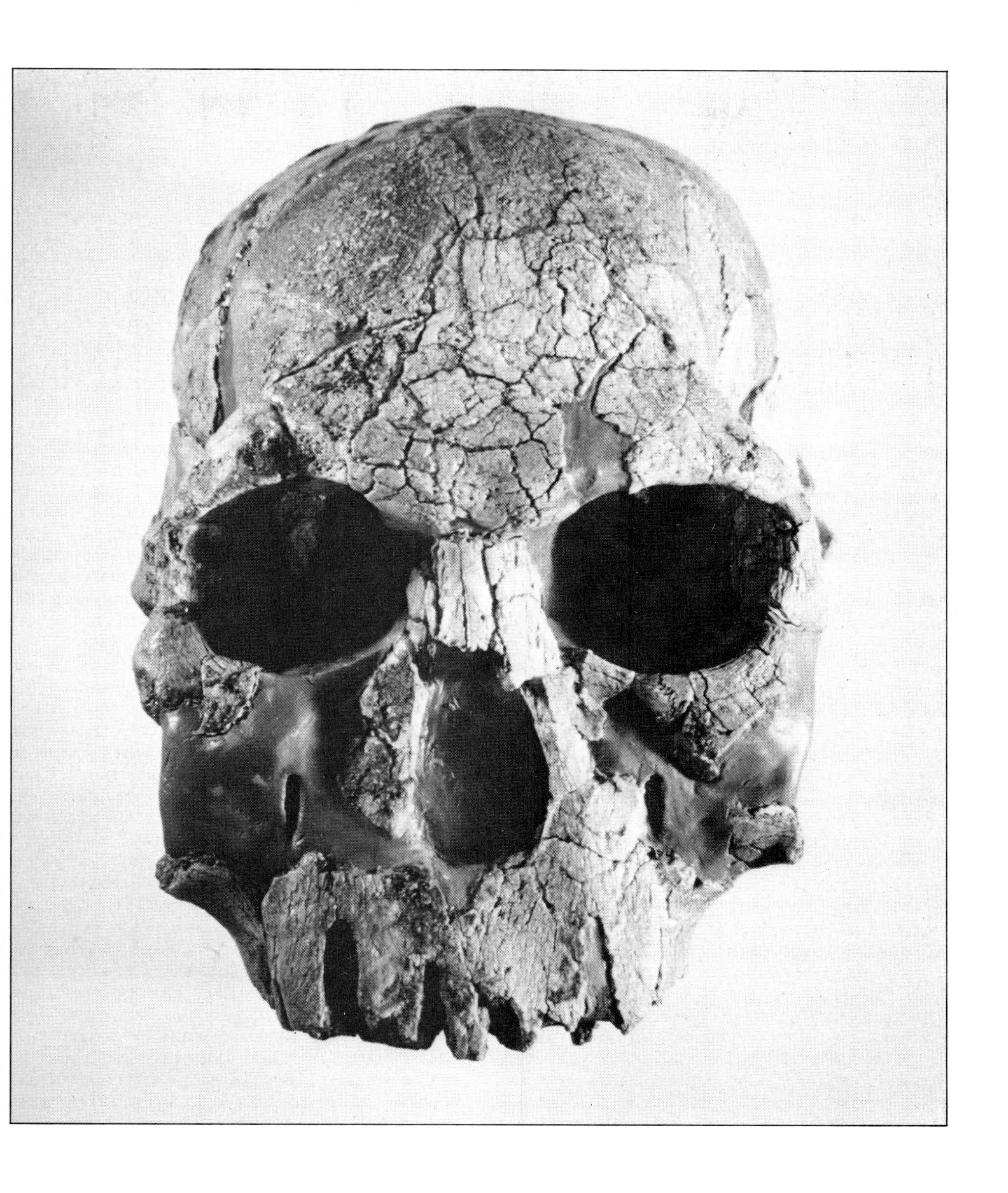

five million years. The theory is that around five or six million years ago the ancestral stock, *Ramapithecus*, suddenly diversified into several different lines, probably because climatic or other environmental changes formed new habitats for exploitation. (Incidentally, there is evidence that other creatures also diversified at this time, adding weight to the notion of a general environmental change.) If this is true, it means that as one approaches closer and closer to the time of speciation it becomes increasingly difficult to distinguish the fossil remains of one hominid type from another: they will tend to all look more and more like the ancestral hominid from which they evolved, and therefore more and more like each other.

In the flood of fossil finds throughout East Africa that have all but overwhelmed us since the portentous 1972 discovery of 1470, this prediction has been born out: spectacular fossils, found particularly by Don Johanson and his colleagues in the arid wastes of eastern Ethiopia, but also by Mary Leakey and her helpers in long-neglected fossil-bearing deposits a few miles from Olduvai Gorge, include examples of primitive *Homo* of nearly four million years old! Just a few years ago no one in their right mind would have believed this possible. But it is, and the evidence is there for all to see – and to wonder at, for it is awe-inspiring to try mentally to grapple with the notion that our direct ancestors inhabited the earth so very long ago. Meanwhile, too, some quite unexpected and important fossils have been discovered and unearthed from the East Turkana deposits, discoveries that give us the most complete and probably the earliest examples of *Homo erectus*, our immediate ancestors who, for a reason we can only guess at, crossed the thin strip of land that joins Africa to Asia, thus beginning mankind's present domination of the world.

But we are running ahead of ourselves, leaving 1470 behind us too soon. For instance, not long after Bernard Ngeneo first spotted those few scraps of fossilized bone that gave us 1470, John Harris, a paleontologist in the team was examining the fossilized remains of an elephant being eroded from the ground, when, in the middle of the shattered pieces, he noticed sections of an almost complete thigh bone (femur) and the top and bottom parts of the lower leg (tibia and fibula) of a remarkably advanced hominid. When they were examined closely there was practically no difference from modern human leg bones. Could this be the leg of 1470, gnawed and dragged away by one of the many carnivorous scavengers with whom our ancestors shared their lakeside habitat? We shall never know for sure, but without doubt it came from an advanced hominid like 1470.

We can be sure, however, that 1470 and his companions could walk with an easy striding gait in the same way as modern humans. In this they may have had an advantage over their australopithecine cousins, because the anatomical arrangements of the leg bones and the pelvis in these creatures appear to have been unsuited to habitual upright walking. This is not to say that the australopithecines dragged themselves around with an awkward stooping gait so beloved of cartoonists of human prehistory. Very probably their special anatomy simply prevented them from walking in the style of their *Homo* cousins. Whatever difference there was, it was certainly not dramatic.

When 1470 and his hominid cousins shared the shores of Lake Turkana the waters were much more extensive than they are today, and the plant and animal life were almost certainly richer too. In fact the history of the lake has been anything but uneventful, rising and falling in erratic cycles through the ages. Less than 10,000 years ago the waters were 200 feet higher than they are now, making the present 150-mile-long lake look like a mere puddle. What caused the dramatic plunge in the water level no one knows, but the presence of the huge Nile perch in the lake is a delicious reminder of the time when Turkana was high enough to be a source of that great river.

The fluctuations in the lake level are not a matter of indifference to fossil hunters because, had the waters remained constant while our hominid ancestors and their cousins played out their lives on and near its shore, the East Turkana team would have pitifully few fossils to search for. Gently-rising waters soon cover bones with fine silt, thus protecting them from the sun, the wind, and most important of all, scavengers.

Carnivorous scavengers are probably the single most significant reason why *complete* skeletons of ancient hominids are never found. This is being dramatically demonstrated by Diane Gifford in a somewhat unusual research project around the fossil sites of East Turkana. She is monitoring the fate of animal corpses – her 'critters' she calls them – over a period of years. Usually the first thing to happen to a dead animal, mostly zebra in her collection, is that it is stripped clean by vultures, hyenas and jackals. But that is not the end of it. The hyenas do not lose interest at this stage: they like to chew the tough cartilage and gnaw the bones, pulling and tugging the skeleton as they feed. They may take bones away so they can savor them in peace or feed their young back at their lair. Meanwhile, herbivorous animals make their impact too, shattering and crushing the larger bones as they trample by. The hooves of these passing herds however, may help to preserve the small bones, by pressing them into soft earth without smashing them.

If you visit a site where an animal has been dead for more than a year, you should not expect to find a neat and tidy skeleton gleaming white in the sun. With luck there will be a skull (often broken), perhaps one or two intact limb bones, the odd vertebra, and hundreds of bone fragments, all scattered over a fairly wide area. Even if everything were collected, there would still not be a complete skeleton because the many bits and pieces that hyenas and jackals – and even porcupines – had taken to their dens would be missing. Such would be the fate of a hominid who died two million years ago. Indeed, until the flame of self-awareness burned bright in the human mind, encouraging him to bury his fellows because of the importance of the spirit and the afterlife, virtually every deceased hominid finished up like Diane Gifford's critters: dismembered, crushed, and scattered.

The process of fossilization is, in detail, still something of a mystery, but in general it depends on the chemistry of the soil and the availability of minerals that can replace those already in the bone. So, not all of the pieces of bone that survive the ravages of the early destruction phases are eventually fossilized, most simply crumble and vanish into the dusts of time. No wonder fossil hunting is a frustrating business! In quiet moments one can fantasize about discovering a prehistoric Pompeii, a group of hominids engulfed in a river of volcanic ash. What a find that would be! For the first time we would be able to construct reliable measures of the relationship of the size of the brain to the rest of the body – at the moment this is one of the crucial aspects of our ancestors' bodies we have to make guesses about. We would be able to assess the degree of individual variation within a group, and for the first time we would have a certain count of the number of individuals in a group. We would see how many men, women and children there were too; and, although they would have been stopped in their tracks, we would see better than ever before what sort of activities were going on in the camp.

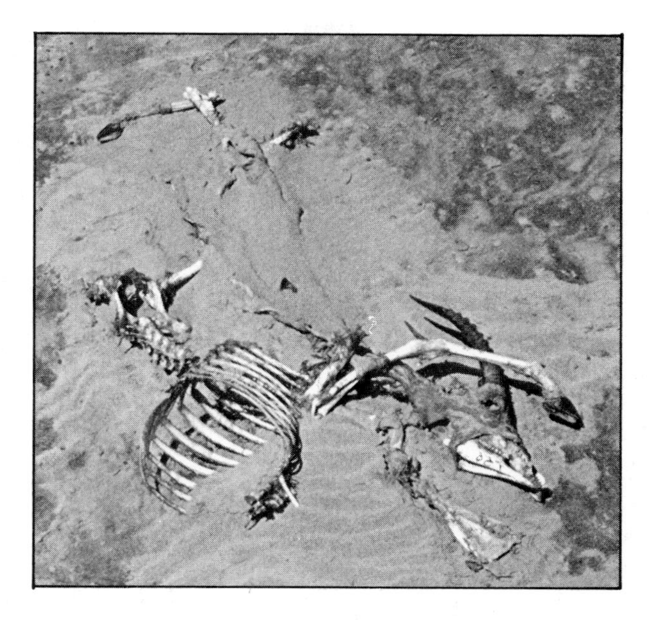

Of no little interest for archeologists are the studies made by Diane Gifford on the fate of animal corpses; above, the remains of a topi, photographed within a month of its death and, below, the scattered remains of a zebra killed by a lion on the lake shore south of Koobi Fora, photographed about eight months after its death.

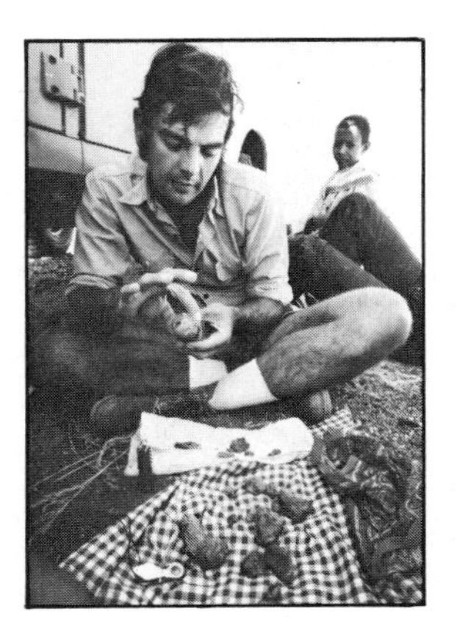

Don C. Johanson (left), of the Cleveland Museum of Natural History, has been working in the Afar Depression in Ethiopia. Below, he is using a small air hammer to chip away the matrix in which is embedded the skull of a child about five years old – one of a family of the genus Homo.

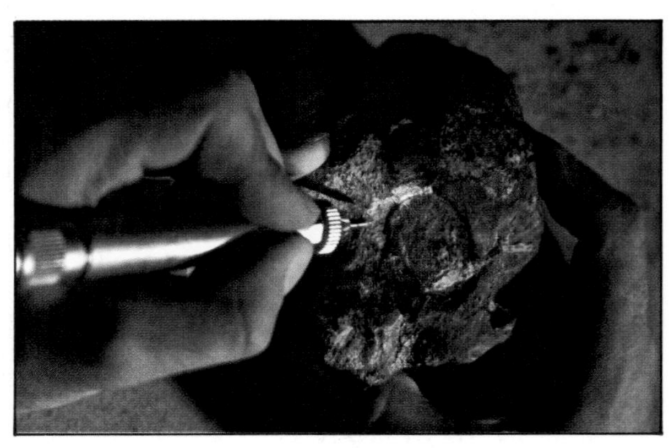

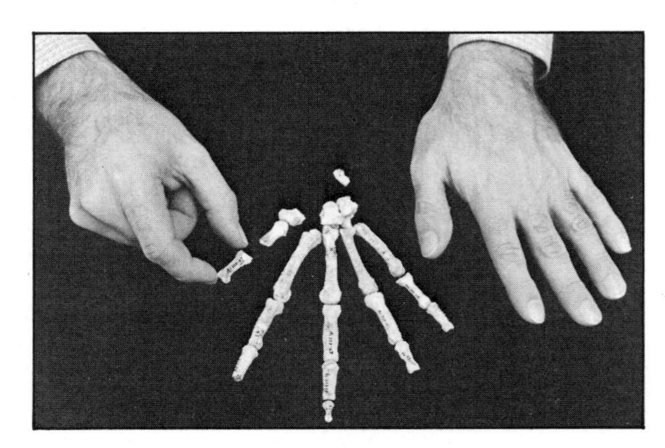

Also found by Don Johanson and his team in the Afar Depression were thirty-five hand and wrist bones – sufficient for him to reconstruct a composite hand shown here with his own hands for comparison.

The fantasy could go on for ever. But these are the sort of questions that we have to try to answer aided by clues so much less complete than those provided by the mythical prehistoric Pompeii. The closest to the fantasy that anyone has so far come is Don Johanson and his team in the Afar. He came across a collection of bones buried some three million years ago. As far as he can guess, and it is no easy task, the bones represent a group of between five and seven people, two of whom were infants of around five years old. Very probably they were one or two family groups. Although Johanson's discovery falls far short of the Pompeii fantasy, it certainly represents the earliest clear fossil evidence of a group of hominids living together. But why did they all die together? What catastrophe overtook them? We now know they did not die in a flood. Perhaps they succumbed to a particularly virulent disease.

Johanson's fossil group is especially important because it belongs to the genus *Homo*. It so happens that in the collection there were enough bones for him to be able to construct a composite hand and much of the skeleton. Although the bones did not all come from one individual, they fitted together well enough for it to be apparent that the size was very similar to modern human hands. (It turns out that the adults were somewhere between four and five feet tall.) But, more importantly, the details of the hand suggest that it would have been almost as dextrous as modern human hands. Remember, we are talking of individuals who lived about three million years ago! When you look at your hands you are seeing structures which, though uniquely human in their manipulative skills, were designed far back in our evolutionary history. If you had been born three and a half million years ago your hands would not have been so very different. With what skill and intent you would have been able to use them, though, is yet another fantasy with which to conjure.

Although, as far as we know, the volcanoes of East Africa have not provided us with the gift of a prehistoric Pompeii, they certainly play a crucial part in our interpretation of the fossil remains. Specifically, they have provided us with a timescale by which we can put an age to the specimens we unearth, and this is possible because, by certain physical chemical tests based on the slow but regular conversion of an isotope of potassium to an isotope of the gas argon, one can determine to within a reasonable degree of accuracy

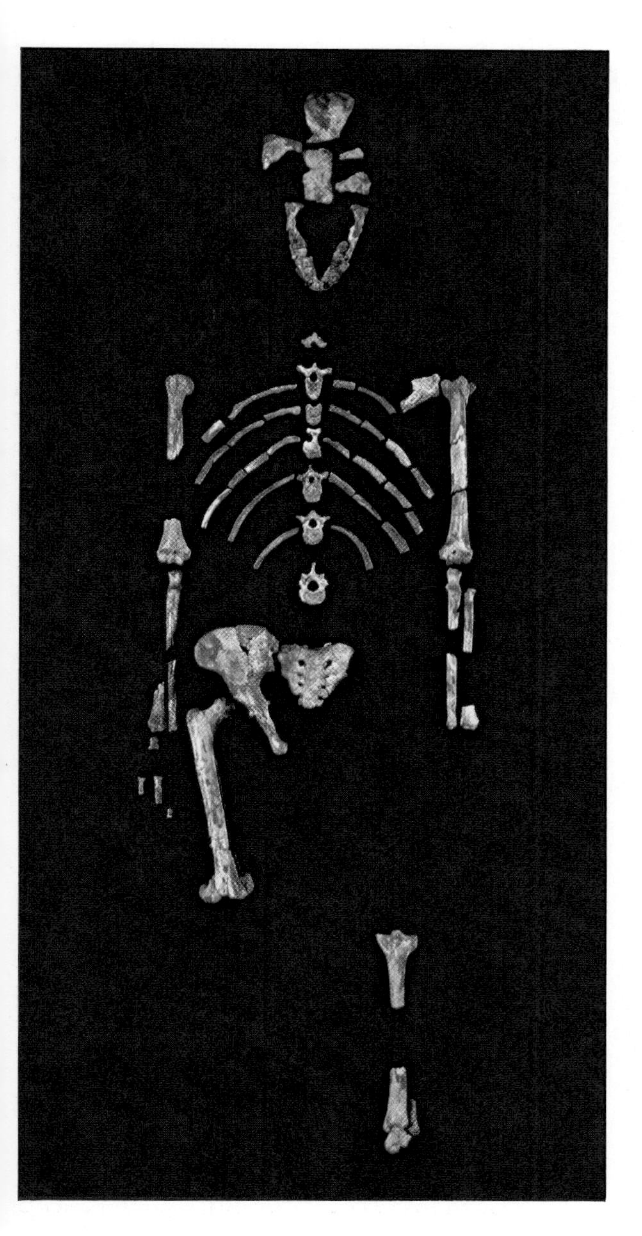

Another very important Johanson find was the near-complete skeleton of 'Lucy' (named after a Beatles' song which was being played on a tape recorder in the camp). The skeleton is possibly a female member of a late form of Ramapithecus. *The inner edge of the pelvic girdle gives the clue to her sex.*

Dating Fossil Hominids

The accurate dating of fossils, or the sediments in which they occur, is essential for a better understanding of our evolution. The best-known dating method is the use of an isotope of carbon (C^{14}), which is accumulated by all living animals through the food chain. When the animals die the accumulation ceases and the C^{14} changes (decays) to an isotope of nitrogen (N^{14}). The rate of decay is known and by comparing the amount of C^{14} with that which was in the bone at death it is possible to date the bone. However, after about 50,000 years so little C^{14} remains that accurate dating is not possible, so other methods have been devised – methods that date the fossiliferous deposits.

Potassium-argon dating

This system uses the same basic principle as the C^{14} method but is based upon the decay of an isotope of potassium (K^{40}) into an isotope of the gas argon (Ar^{40}) in volcanic deposits, particularly the ash (tuffs). Because the rate of decay of K^{40} is slower than that of C^{14} this method extends our dating techniques to cover the range of human evolution.

Fission-track dating

This comparatively new method was devised after it had been observed that the explosions occurring during the decay of U^{238} (an isotope of uranium) in glass produced fission tracks that could, using a suitable acid for etching, be made visible under a microscope. Exactly the same happens in the natural glass of volcanic deposits. The tracks in the specimen to be dated are first counted; the specimen is then placed in a nuclear reactor to cause the balance of the U^{238} to undergo fission; the new tracks so formed are then counted, and the sum of the two counts will give the total of U^{238} that was originally in the specimen. As the rate of decay of U^{238} is known, the date of deposition can be calculated. The dates so obtained can be compared with those found from the potassium-argon method and, because the experimental errors in the two methods are different, they form a useful check. In the Olduvai Gorge area the dates do show a fair measure of agreement.

Geomagnetic dating

In sediments that do not contain sufficient potassium or uranium other methods of dating must be devised. One such uses the fact that during the earth's history the magnetic field has changed direction. When being deposited the metals in the sediments will take up the direction of the field then prevailing and thus the magnetic field becomes 'fossilized'. This provides a separate time chart for dating deposits which can serve independently or as a check against other systems.

the time at which the ash was hurled out of a volcano.

What happens is this. Over the very long periods of time that cover human prehistory, silt carried either by lakes or by streams, and wind-blown sand builds up to form deep layers, layers that eventually turn into rock. Naturally, these sediments are older at the bottom than at the top, having been laid down first. The rate at which they are formed, and therefore the final thickness of the sediments, is determined by the amount of silt and sand available. For instance, at Olduvai Gorge there is roughly 300 feet of such sediment and ash, and this was formed over a period of about two million years. In the Omo region of Ethiopia, however, sediments half a kilometer thick were deposited during the same period of time. It is within these deposits that the hominid bones become entombed and fossilized. But as it is impossible to date either fossils or non-volcanic rocks directly, we can only establish that one fossil is older than another if it is situated lower down in the deposits. This is why volcanoes, or rather the ash from them, are so important.

When volcanoes erupt they often send out clouds of ash which eventually settle on the ground, either directly as an ashfall, or after being carried and deposited by flowing water. These layers of ash simply form an extra layer in the steadily-growing strata of deposits, but each layer is special because it carries a potential date stamp of the eruption. And with a series of such dates spread through the deep deposits one has a rough timescale by which to measure the fossils. For example, if a fossilized animal is found below a layer of volcanic ash that is two million years old, but above the next lowest ash layer that was deposited three million years ago, then it must have lived between these dates. As a general rule the nearer it is to the upper layer of ash, the closer the age will be to two million years, and so on.

This system is proving invaluable in the major East African fossil sites, in Ethiopia (at Hadar and Omo), east of Lake Turkana in Kenya, and in Olduvai Gorge and Laetolil in Tanzania. Although the dating method is not without problems, it does provide the crucial time framework within which hominids from different areas can be compared: it is interesting to know, for instance, if hominid evolution is pushing ahead more speedily for some reason in one area rather than another. And one might be able to spot pockets of hominids that have been 'left behind' in the march to mankind.

An excavation at East Turkana, where stone implements were recovered.

One thing we can be sure about is that the emergence of human ancestors was a dynamic affair: there was neither a uniform trend throughout the hominid populations of Africa, nor was there a single location where some kind of magic step was taken, producing a 'master race' that seeded the African continent, and ultimately the rest of the world. Evolutionary spurts influenced by local conditions produced local variants, this much we know. For instance, the robust australopithecines that lived along the side of Lake Turkana were substantially stockier individuals than those that lived in southern Africa. Similar kinds of variation in build and size developed among the *Homo erectus* populations spread throughout Europe, Asia, and Africa a million and more years ago. You only have to look around the world today to see that, although all humans are undisputably members of the same species, there are distinct geographical characteristics.

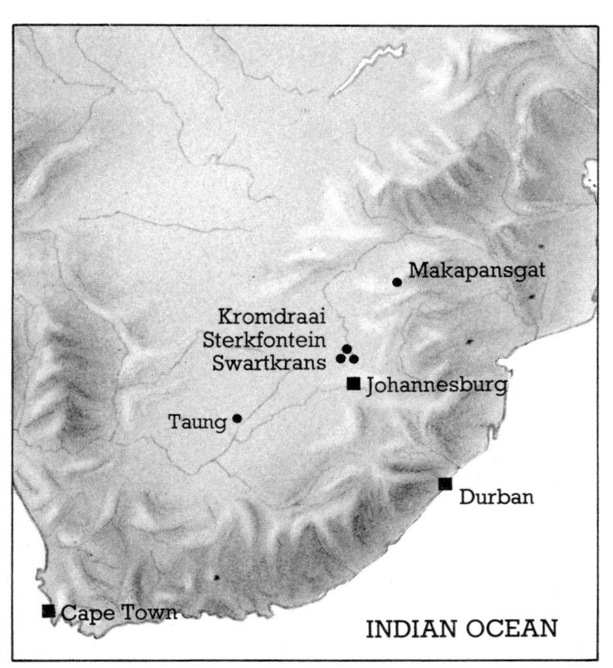

The southern African sites which have yielded clues to the evolution of the australopithecines.

Mention of the southern African australopithecines takes us right to the birthplace of these creatures, in terms of fossil hunting that is, not their biological emergence. Late in 1924 a miners' explosion in a quarry at a place called Taung ('the place of the lion' in Bantu) broke away a rock containing the fossilized skull of an infant who had died a very long time ago. The rock was one of many that were crated up and sent to Raymond Dart, Professor of Anatomy at the University of Witwatersrand. The skull of an ancient baboon and other bones had already been found in the debris of the quarry, and Dart was curious to see what else might turn up. The Taung baby, as it was christened, was more than he could have hoped for; it was certainly more than the world's prehistorians, which meant principally those in Britain, France and Germany, could take seriously.

Dart discovered that the skull was relatively large compared with other known non-human primates; the teeth were more hominid-like than ape-like, as was the shape of the face. Judging by the estimated angle at which the head joined with the neck, the creature

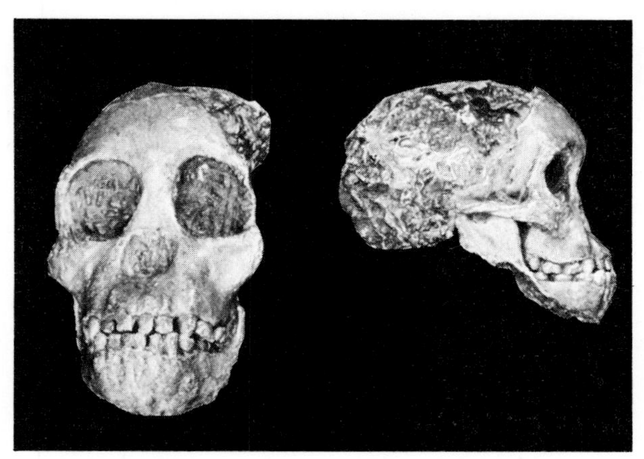

Two views of the skull of the Taung baby – a five-to-six-year-old Australopithecus africanus. *This fossil had been found in a limestone quarry near Taung in Botswana and sent to Raymond Dart, at the University of Witwatersrand in 1924, who immediately recognized its great importance.*

stood and moved in an upright posture. It was clear to Dart that this was a human ancestor, and he declared so in the scientific journal, *Nature*, at the beginning of 1925. At the time much of the scientific world had been seduced by the Piltdown forgery, a combination of a modern cranium with an ape's jaw which some still unknown hoaxer placed in a gravel pit in Sussex, England. This 'discovery' in 1912 of the unnatural hybrid fitted very well the notion that human ancestors must have had a large brain, and that the upgrading of the rest of the body from ape-like to man-like form trailed behind. In the face of this firm belief, backed up by the Piltdown evidence, Dart's discovery was largely ignored, and even scorned: it may have been an ancestor of an ape, people conceded, but certainly not a human.

Undismayed, Dart worked on, and before long others, including Robert Broom, John Robinson, and later, Philip Tobias, joined in the search in several parts of southern Africa. Within twenty years 'apemen' had been discovered at four other sites, together with some evidence of stone tools. At three of the sites, Taung, Sterkfontein, and Makapansgat, the hominids were all quite similar to Dart's original discovery: they were the gracile australopithecines, named by Dart *Australopithecus africanus*. At the other two sites, however, a large and more stocky type had lived, and this had been named *Australopithecus robustus* (the South African form of *Australopithecus boisei* in our original list).

The fossil collection from these sites is impressive, and it contains many bits and pieces of limb and other bones, including parts of a pelvis for both *africanus* and *robustus*. The thigh bone shows a curious difference from *Homo* thigh bones; at the joint with the pelvis the 'ball' is smaller than in *Homo*, and the 'neck' to which it is attached is longer and flatter. These

Left: The Great Rift Valley of East Africa is one of the areas on the earth that are still geologically unstable. It is a remarkable feature of the earth's surface – a continuous groove that extends for more than 2,000 miles, dotted with volcanic cones and craters. The volcanic material provides invaluable clues that enable geologists to date the deposits in which fossils are found. The photograph shows the plateau scarp at Lake Manyara, Tanzania.

Right: A reconstruction of Australopithecus africanus.

and other characteristics have been interpreted to mean less effective walking in the australopithecines. Nevertheless, these creatures were quite definitely habitual upright walkers and there is no evidence to support the suggestion that they would occasionally pound along on all fours in the knuckle-walking style of chimpanzees or gorillas.

At all but the Taung sites, particularly in the younger deposits, excavators have found stones that appear to have been shaped purposely as tools. Most are crude and may have been used for crushing nuts or preparing other plant foods, or breaking bones to reach the succulent marrow. There are so-called scrapers too, flatter tools with edges that might have been used for scraping hides or stripping bark; but it is only a guess. One of the problems in describing tools, or rather artifacts as they are more often called, is that even the simplest implements can be used in many ways. But an even trickier problem is, who made the tools? Again, if we are honest, we have to admit that we shall never really know.

It is reasonable to say that *Homo* was a tool maker because this is seemingly a part of the package that led to the emergence of mankind. But what about his australopithecine cousins? Unless our prehistoric Pompeii were to yield a squatting *Australopithecus africanus* caught in mid-swing while fashioning a stone implement we shall never be able to say, yes, they were tool makers too. It is well known that apes can be taught to use tools, and even make crude cutters, as Abang, the Bristol Zoo orangutan, demonstrates. It would therefore be surprising if the australopithecines did not occasionally use stone and wooden objects as tools. But whether this hominid species manufactured tools to particular patterns, creating the sort of stone cultures we know existed at least two million years ago, is quite a different question. Their way of life may well have made no demand for structured tool kits, though the odd bespoke stone cobble might have been useful.

One idea of a tool industry which has now been scientifically discounted, but which lingers on in the popular image of 'ape-men', was borne out of the discoveries at Makapansgat. Like most of the South African fossil sites, Makapansgat is a cave collection and was almost certainly a place where dead animals, including hominids, finished up, rather than being a real occupation site or home base. Raymond Dart was intrigued by the curious assortment of animal bones that were found together with the remains of hominids (the gracile *Australopithecus*) and he speculated that these bones were part of what he called the osteodontokeratic culture (literally, bone, tooth and hair culture). He suggested that *Australopithecus*'s tool kit included teeth-filled jaws which could be deployed as serrated saws, limb bones used as clubs, and other more fanciful notions. This tool kit was purported to be part of our ancestors' alleged carnivorous and cannibalistic way of life.

But recently, in a brilliant piece of scientific detective work, Bob Brain effectively demolished the notion of an osteodontokeratic culture. In a series of carefully-planned experiments Brain monitored the effect of both weathering and scavengers on collections of animal bones. Over a period of years Brain observed that a combination of scavenging habits of the local carnivores, and the differential resistance to weathering of various types of bone, produces a bone collection virtually identical to the one Dart found in the cave: the osteodontokeratic culture is apparently no more than the left overs from many leopard and hyena meals!

This cautionary tale emphasizes the problems of interpreting the material found on excavations, and it points to the value of the apparently morbid business of studying the moldering animal bones – like Diane Gifford's critters – so as to discover what might happen between the time of death of one of our ancestors and the time the remains feel the gentle stroke of the archeologist's brush.

When out fossil hunting, it is very easy to forget that, rather than telling you *how* the creatures *lived*, the remains you find indicate only *where* they became *fossilized*. This important caveat applied particularly to the South African fossils, which were mainly recovered from limestone caves. They may have fallen into the caves, to become part of a conglomerate fossil assemblage, or been taken in by carnivores, either having been hunted and killed, or simply chanced upon as a piece of dead fresh meat. The five main sites in southern Africa, though a valuable source of fossils, are mainly scavenger lairs rather than hominid living sites. The home bases, if they existed, have not been discovered. And, apart from a recent report by Philip Tobias for the recovery of a partial cranium in the fall of 1976, no contenders for early *Homo* have turned up either. The chances are that some of the species were living in the area, having made the tools, but they seem to have managed to escape the fate of having their bones dumped in caves.

Unfortunately, the caves not only do not reflect a picture of the lives of the hominids, but they also have the frustrating problem of being virtually undatable. There is presently no certain way of saying accurately when a particular skeleton came to rest in its particular cave. Without the benefit of volcanic date stamps, researchers have had to infer geological age by looking at the animal fossils associated with the hominids: this technique, known as faunal correlation, involves the comparison of evolutionary stages and is notoriously capricious – sometimes it can be quite accurate, sometimes not. By these criteria, and some other studies, the hominid levels in the caves have been said to be between more than one million years and less than three million years, with Kromdrai being the youngest and Makapansgat the oldest. These dates cannot be thought of as firm at the moment, although ultimately it may be possible to piece together a more reliable time scale by looking in the rocks for signs of specific times in the past when the earth's magnetic field swung round, an event that has occurred at known times during the past four million years.

If we travel north again, as far as the south-east edge of the vast Serengeti plain in Tanzania, we find a complex of extinct volcanoes surrounded by land which is today quite dry for most of the year. But about two million years ago, under the shadow of this great volcano complex was a lake fed by innumerable streams draining from the highlands. Animals came to drink at the lake edge. And it is certain too that these animals shared the lake shore with bands of hominids.

From about two million years ago, lake sediment and wind blown sand steadily built up and up, until now that distant past is buried beneath some three hundred feet of deposits. Earth movement associated with the geological stirrings in the Great Rift Valley drained the lake. Ultimately, a quirk of nature, in the form of a seasonal river, has carved out a gorge, slicing through the deposits of sediment so that now you can stand on the bedrock and gaze at millennium after millennium stacked neatly as a layer cake of time. And the critical layers of volcanic ash have been dated, providing an invaluable time scale. This is Olduvai Gorge, a 25-mile-long gash in the arid plains containing a rich array of human prehistory.

Long before any hominid fossils were discovered in the Gorge's ancient dry sediments, Louis and Mary

The late Louis S. B. Leakey, who spent nearly four decades searching for prehistoric life in East Africa. He is shown here with the broken molar of a Dinotherium *in his hand and, on his hat, the tooth of a million-year-old elephant found at Olduvai Gorge. Below, his wife, Mary, herself a distinguished archeologist, working in the Gorge.*

Leakey were finding what is now established as the most complete perspective of stone-tool cultures of this important period of human evolution. Since the mid-1930s Mary Leakey, particularly, has been analyzing the tool industries on more than twenty 'occupation' sites scattered in a time range of between one and two million years ago. As one might expect, over that huge track of time there is an increase in the sophistication of the tool technology, but the progress is very gradual indeed, and there is a curious overlap of apparently different cultures.

In the lowest layers of the Gorge sediments, Bed I, the tools, quite naturally, are crude, made by knocking a couple of flakes off a tennis-ball sized pebble. The result is a chopper, the central piece of equipment in the so-called Oldowan industry. Crude scrapers and possible hammerstones also make up part of the tool kit. For a million years this culture continued readily identifiable at Olduvai, but with just enough embellishment on the basic pattern towards the end to justify calling it the Developed Oldowan industry. A million years ago this tool technology disappeared at Olduvai.

However evidence of a distinctive second tool culture was left along the ancient lake shore, and this is known as the Acheulian industry. The hallmark of this industry is a very characteristic, but also rather mysterious object known as a hand ax. Shaped like a large tear-drop with sharpened edges, the hand ax is clearly an object that requires considerable skill and preconception for its manufacture, but what it may be used for is anyone's guess: hacking wood, smashing bones, anything is possible, but none seems to require the specific shape that must have demanded great investment of effort and time to create. In fact, any group of prehistorians sitting chatting after dinner is likely to stray onto the topic of the function of hand axes, to abandon it later having generated yet another set of wild and entertaining guesses, but having no evidence on which to test any of the suggestions. The practitioners of the Acheulian industry, apparently *Homo erectus*, quite clearly had an intricate material culture because in addition to the hand axes, they used well designed choppers, chisels, scrapers, cleavers, awls, anvils, and hammerstones, as well as several variations on these. In no sense was this tool kit the result of random stone knapping; there is a well recognized pattern, just as there was, albeit to a somewhat lesser degree, in the Oldowan industry.

It is not surprising that a more complex and ordered technology should follow a lesser developed one. But what *is* surprising is that, undisputably, the two cultures coexisted for a very long time. The Developed Oldowan appears to bow out at around one million years ago, by which time the Acheulian culture had been established in the area for at least half a million years!

For half a million years two relatively sophisticated stone-tool technologies thrived side by side, and yet they remained distinct. This is a staggering thought. The people producing these two separate technologies shared the same lake shore; they shared the same plant and animal resources; they must have seen each other, and been aware of each other. What does it mean?

The fossil-bearing sites in East Africa (see map) have provided a rich array of hominid remains. Olduvai Gorge (above), made famous by the discoveries of Drs Louis and Mary Leakey, is about 300 feet deep and some twenty-five miles long. It was cut by a river through ancient deposits, now laid bare, originally laid down in a lake. A cross-section of the exposed strata is shown right.

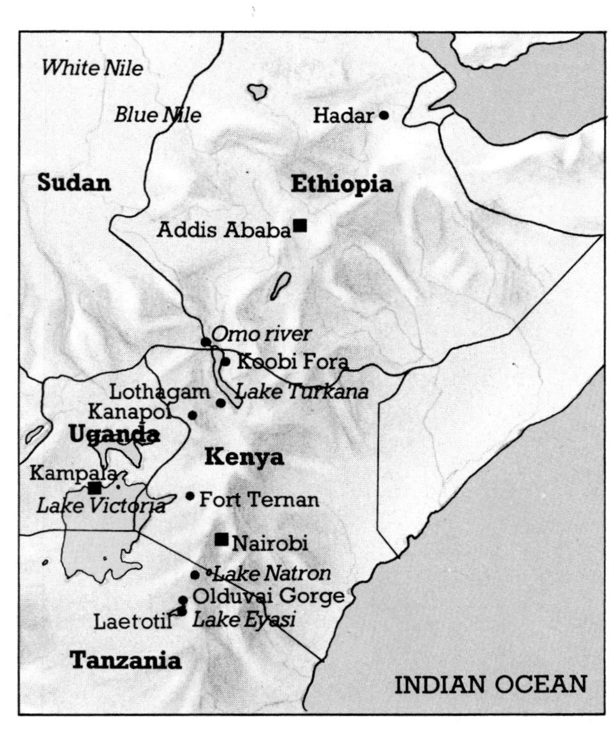

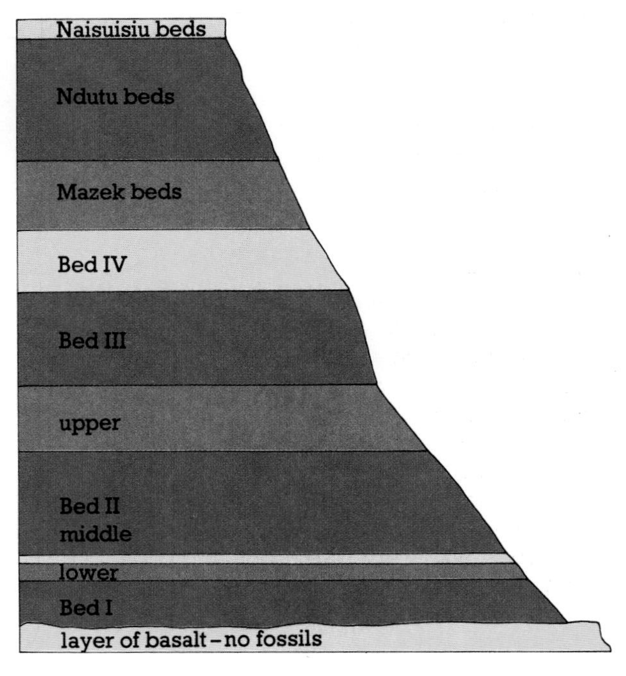

The drawing of the primitive chopper, above, is an example of a tool from Bed I of the Olduvai deposits. This Oldowan industry extends into Bed II, the tools improving sufficiently to justify the use of the term Developed Oldowan Industry (center). Also in Bed II are the tools of a more advanced technology – the Acheulian; the example drawn on the right is a hand ax.

There are, of course, many possible explanations. At one extreme is the notion that between two and one and a half million years ago the hominids inhabiting the lake shore lived lives undisturbed by outside influence, very gradually increasing the sophistication of their tool technology. Then some newcomers arrived who already practised a more advanced technology, the Acheulian industry. By definition, the newcomers would have been *Homo erectus*. The scenario continues, suggesting that the immigrants and the indigenes lived in close proximity, but with little significant conflict, until the less-advanced hominids gradually succumbed to the everyday competition for resources. An alternative suggestion saves the Olduvai indigenes from evolutionary oblivion and has them interbreeding with the 'outsiders' until the two gene pools are completely mixed, but again with the superior technology prevailing. This sort of interaction must have happened many times during the course of human evolution.

At the other extreme is the idea that there was no 'invasion' and that what we are seeing in the archeological record is merely that transition of one type of subsistence economy (associated with the Oldowan technology) to another (requiring the Acheulian), and that during the transition both economies are practiced. The Olduvai hominids may have occupied a camp for a few days or weeks and manufactured Oldowan tools, the theory goes, and then moved to a different area where they switched to a different subsistence method, and hence a different set of tools was required. If, for instance, a band of hominids were to spend some of their time on the plains gathering fruit, nuts, and other plant foods, and catching and scavenging the occasional animal, and the rest of the time on the lake shore where they might have lived more on turtles and catfish, then the two ways of life would indeed be separate. But whether this would imply separate, patterned, and complex tool technologies is rather doubtful.

The fact that two peoples may have similar economies, share the same land, and yet differ culturally in several ways, is illustrated by the Tugen (left) and Njemps (above), who live near Lake Baringo, Kenya. Although their villages are in close proximity and they have social and trade contacts, they differ in many ways, for example, their styles of self-adornment.

The *style* of material technology in contemporary 'primitive' peoples is determined not so much by the jobs that have to be done as by a need for group identity, for conforming to the tribe's customs. Clearly, the functional context of the technology provides a basic blueprint for the technology, but the way in which, say, a cutting edge is produced and used is determined by arbitrary but consistent rules of the group. Self-adornment, of course, takes this much further and is often a very powerful expression of group identity. Is this relevant to the two cultures at Olduvai?

Obviously the relevance is not total because two-million-year-old hominids had less capacity for creating cultural identity than do modern humans, but the experience of British archeologist Ian Hodder is certainly salutary in this respect. He is studying cultural and material patterns in two tribes, the Tugen and the Njemps, who live among the low rolling hills near to Lake Baringo in Kenya. Both tribes have similar economies, based on cattle, sheep and goats, though the Njemps are just slightly more inclined towards a pastoral existence than their neighbors. Nevertheless, they share the same land and their villages are very close together. In spite of their proximity, a proximity that goes back over many years, the two peoples are very different in their style of self-adornment, material products, and in the structure and arrangement of their houses. Moreover, they speak different, though related, languages.

Such clear definition between the two groups would be understandable to some extent if there were no social or trade contacts. But there are. The two tribes have traded goods between each other for many years. And a number of Tugen women married Njemps men, whereupon they assumed Njemps decoration and made baskets and other goods in the Njemps style.

The reason for the sharp differences in material culture and decoration, Hodder believes, is that close identity with the tribal group is necessary for maintenance of the social structure. There are firm rules of kinship interactions and marriage, and without a strong sense of conformity these would break down. So, if an archeologist were to excavate abandoned camps of the Njemps and the Tugen, the differences in material pattern he would find would reflect not differences in economies but rather an expression of belonging to a particular social group.

Back at Olduvai, where one and a half million years ago social life and organization were undoubtedly simpler than those of modern humans, there is nevertheless the real possibility that a self-awareness and a feeling for form and shape might have influenced the way tools were made. An Acheulian hand ax is a beautiful piece of work. It is surely not stretching the bounds of possibility too far to suggest that its maker would not have been totally unaware of that fact. We know that the makers of the Developed Oldowan and Acheulian industries had a firm conception of the implements they were manufacturing, simply because of the consistent patterns within the assemblies. Very probably at least part of that conception was governed by matters of style as well as utility. We are not suggesting that the execution of the style in each implement is part of a conscious fancy for a particular form. It is much more likely at this early stage of cultural expression to be simply an aspect of familiarity with a specific environment, and the finished tools are part of that environment. In spite of the undeniable sophistication of the Acheulian tools compared with the Developed Oldowan, there are not many jobs that can be done exclusively with Acheulian implements.

All this points to the possibility at least that the two cultures at Olduvai, although they are part of a steady sophistication, may have been produced by different 'tribes' of the same indigenous lakeside population. The aggregation of animals of the same species into groups with which they have a strong attachment is, of course, not unusual in the animal kingdom. But what *is* special is that the sense of belonging to a certain group is expressed in the style in which inanimate objects are fashioned. It is fascinating to speculate, but we must not push the speculation too far, that we are seeing the first glimmerings of tribal expression in the artifacts at Olduvai.

While the Olduvai hominids, whoever they were, were thriving one and a half million years ago on the game-rich shores of the fluctuating lake and were creating their two distinct stone industries, human ancestors on the eastern shore of Lake Turkana were adding to our understanding of the complexity of stone-tool technology by striking yet another stone tool culture, the Karari industry. In an environment not unlike the lake margin, stream-side setting chosen by the hominids who occupied the KBS 'fig leaf' site perhaps a million years earlier, these more advanced hominids made stone artifacts rather like those of the Developed Oldowan industry. Perhaps not surprisingly, the KBS inhabitants had manufactured the more primitive Oldowan-like artifacts. The point is, however, that although the Karari industry is basically similar to the Developed Oldowan, there are differences important enough for it to be classified separately.

For instance, the ratio of the various implements is chacteristically different at the Karari sites, there being an unusual emphasis on heavy scrapers. And a curious idiosyncrasy was the fine flaking of light-duty scrapers which produced a serrated edge to the tool. Very likely the Karari hominids were pursuing different forms of subsistence economies from their cousins at Olduvai, though the differences are unlikely to have been substantial as they both occupied comparable environments offering similar resources. Different economies might account for some of the characteristics of the industries, but again we can guess that local tradition also played its part in shaping those characteristics.

Before the Karari industry became established, the hominids in the East Turkana localities were producing the more or less basic Oldowan technology, using the lava rolled smooth in the streams. If, two million years ago, one of these hominids had decided one day to undertake a journey of some hundred miles or so northwards to the flood plains of the great meandering Omo River of southern Ethiopia that now pours its orange-brown silt-laden waters into the great green Lake Turkana, he would have had something of a shock. Although the hominid populations along the river banks and in the surrounding countryside were not greatly different from the mixture of communities he had left behind on the lake shore and in the nearby hills, the tool technology would be quite unlike anything he had ever seen.

Opposite: The site of the fossil-rich beds in the Omo River Valley, Ethiopia. It was here that Clark Howell and Yves Coppens found quartz 'tools', one of which is illustrated on the left.

In what is now known as the Shungura Formation, Clark Howell, Yves Coppens and their colleagues have found a number of occupation sites containing what appear to be sharp, randomly angular fragments of quartz. Apart from two fragments of lava, which are also probably artifacts, the quartz 'tools' are all that there is to be found in the silty clay of the sites. To call these fragments tools may seem to be an unwarranted presumption, because they certainly do not fall into neat categories of implements we are used to seeing in an industry. But there is no doubt that the quartz was taken there by hominids, because the nearest source of the material is at least several kilometers away; the quartz had been transported to this point in streams running from the primary outcrops which are at least 20 kilometers distant. Once at the camp the blocks of milky-white vein quartz appear simply to have been shattered, producing the fragments that the French and American researchers are now finding at the Omo sites.

Does this reflect the gross unsophistication of a backward hominid? Almost certainly not, because although the quartz, apart from an occasional piece of larva, is the only rock in the flood plain area that can be made into cutting implements, the physical properties of the rock just do not allow the careful shaping that we see at other hominid campsites by Lake Turkana and at Olduvai. Presumably the quartz 'technology' served the hominids well enough. And the choice of the hominids to live by the river and make do with the inferior tool-making material rather than go tens of kilometers into the hills where they would have found enough lava for 'good' stone tools, emphasizes the propensity for water-side living in hominids at this stage of their evolution. Once someone invented a way of carrying and storing water, perhaps in animal skins or in the egg shells of the very large ostrich-like birds that lived at that time, the age-long dependence on water-side living would be broken, freeing our ancestors to explore more extensively the land of their birth.

The story of the Omo hominids tells us that there is yet one more factor that we need to consider when interpreting stone-tool industries at this early stage of human prehistory – the nature of the material available. Utility, cultural tradition, and raw material, therefore, all play their part in generating the heterogeneous mixture of stone-tool industries. Much later in the story of human evolution, after about one hundred and fifty thousand years ago, the mixture of technologies becomes much richer, and there are cultural transitions similar to the one that happened at Olduvai. The difference is that by this time the pace of progress had speeded up. Cultural expression through tool technology was beginning to have a real impact, though again against a background of technological advance. So, when we talk of stone tool cultures, even as far back as the Olduvai and Turkana hominids, we really do mean *culture*, not just *technology*

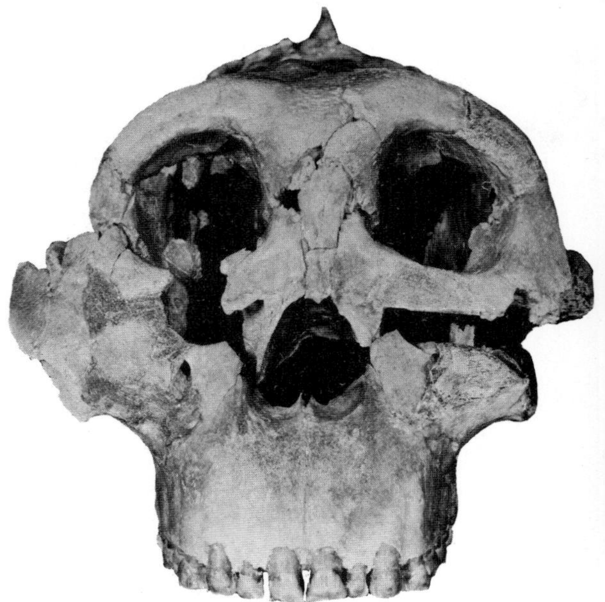

Above: The partially reassembled skull of Australopithecus boisei, *discovered by Mary Leakey in Olduvai Gorge.*

The years between the 1930s and 1959 were both rewarding and frustrating for Louis and Mary Leakey in their search through time at Olduvai Gorge. Rewarding because of the wealth of stone tools they found and the cultural or technological entities that they implied. But frustrating too, because for all those years there was no trace of the hand that made the tools (and this search was not helped by two rhinos with whom they had to share their source of water and who insisted on wallowing in it, a habit that did not improve the taste!) The first important Olduvai hominid was found on 17 July 1959. While Louis Leakey was resting in camp because of a dose of 'flu, Mary Leakey was out scouring some sediments near to where, twenty-eight years earlier, Louis had found some stone tools on his first visit to the Gorge. Following the Leakey tradition that you look and look again until you find what you know must be there, Mary Leakey was rewarded that day with the sight of a few skull fragments. Careful probing and excavation unearthed most of a skull of a robust australopithecine.

Left: A reconstruction of Australopithecus boisei.

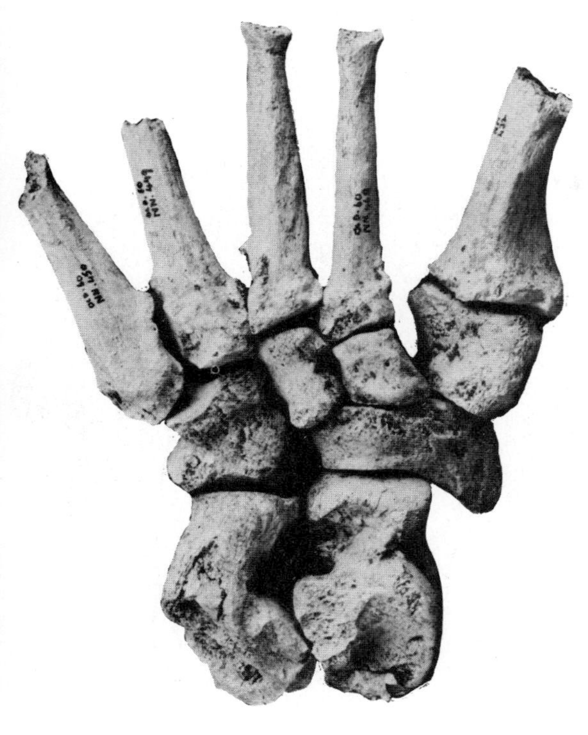

The second type of hominid found in Olduvai Gorge has been called Homo habilis. *Above, fossil foot bones, the anatomy of which suggests that this species could have walked upright.*

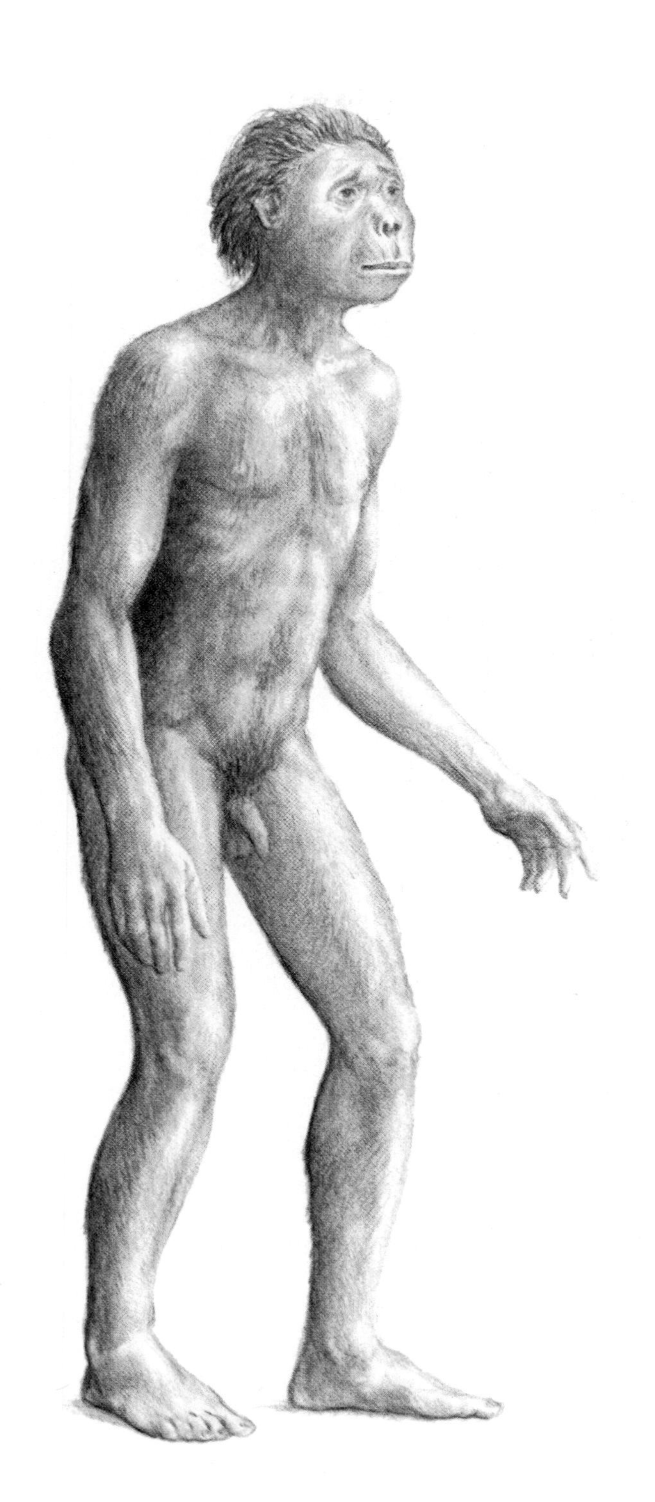

Like 1470, this cranium had been broken into hundreds of pieces, and it required patient assembly to recreate a heavy buttressed skull which, one and three-quarter million years before, had contained a brain slightly more than a third the size of a modern human brain (530cc compared with 1400cc). This creature was similar to the South African *Australopithecus robustus*, but was even bigger. He was named *Zinjanthropus boisei*, East African man, but is now more properly called *Australopithecus boisei*.

Within two years the Gorge had yielded its second hominid, a smaller creature than *boisei*, but with a bigger brain (it is not possible to be certain of the capacity, but it is somewhere between 650cc and 700cc). Because of the size of the brain, and because the teeth were much more human (the molars not being massive in proportion to the incisors), the new hominid was considered to be *Homo*, and in 1964 was given the specific name *Homo habilis* ('handy man'). Here, apparently, was the Gorge's tool maker, or so it was supposed.

Right: A reconstruction of Homo habilis.

Since that time the hominid collection at Olduvai Gorge has grown substantially, and now contains representatives of *Australopithecus boisei, Australopithecus africanus*, and *Homo habilis*, all of which occupied the lake shore simultaneously at some time in its history. This is the emerging scenario for hominid evolution, in East Africa at least, and thus it is not unreasonable to predict that the presence of early *Homo* will one day be confirmed in southern Africa.

The discoveries of fossil hominids in Africa since 1924, and particularly within the last few years, give us the essential basic guidance in our search for our origins. When 1470 made its timely appearance in 1972, it helped galvanize a theory of human evolution that had been implied by the evidence from Olduvai; the theory has subsequently been strengthened by other important discoveries. We now know that if, by using a time machine, we could change places with one of our ancestors living, say, two or three million years ago we would be sharing the countryside with at least two and possibly three hominid cousins (*Australopithecus boisei, Australopithecus africanus*, and perhaps some surviving members of late *Ramapithecus*). We know too that even at that great distance of three million years, the ancestral human stock of *Homo* would have already had a history almost as long. Because of the very nature of the gradualness of biological change we have to concede that, as we move closer and closer to the origin of the *Homo* line, we are faced with increasing problems in interpreting the fossils we might find: during the first throes of diversification the newly-emerging species will inevitably look very much like each other, and like the basic *Ramapithecus* stock.

But we can now point with some confidence to a period in history of about five or six million years ago and say, this is when the human ancestral stock was born. The birth was not a dramatic event, happening just once in one locality. Propitious environmental circumstances (of whose nature we are still ignorant) allowed descendants of *Ramapithecus* to adapt to particular niches within that environment. We know that at least three new forms of hominid (the two australopithecines and *Homo*) emerged during this period. And we can guess that wherever the ecological circumstances were appropriate in various parts of Africa, there *Ramapithecus* evolved into one or more of the new hominids, a factor that must have contributed to geographical variations within the species. The newly-emerging hominids must have spread out from their points of origin, but migration cannot be used to argue for a single, once only birthplace for mankind in Africa. At how many localities *Ramapithecus* slowly evolved into *Australopithecus africanus, Australopithecus robustus,* and *Homo*, we shall never know; but we can be confident that it was more than one.

Once the speciation of *Ramapithecus* into the new hominids was complete, say around four million years ago, there followed a long period of coexistence for *Australopithecus africanus, Australopithecus robustus*, and *Homo* (and also for a while the population remnants of *Ramapithecus*). Ultimately the *Homo* lineage dominated and the australopithecines went into eclipse.

In 1975 a member of the East Turkana fossil-hunting team spotted what turned out to be the oldest and most complete cranium of *Homo erectus* yet to be found in Africa. (Incidentally, it may also be the oldest in the world, but that is still a matter of conjecture.) This large-brained ancestor of ours (it has a brain size of about 900 cc) was living alongside the lake about one and a half million years ago. The discovery of this superb cranium emphasizes very clearly the notion of coexistence, especially so because not very far away the skull of a robust australopithecine had been discovered some time before. What we want to know now is how they coexisted. What were they doing as they lived out their separate lives on the margins of the lake?

This is certainly one of the most difficult questions to answer in human prehistory. All we have to go on is the bones and stones that are strewn along the path of human evolution. And just as litter often finishes up in litter bins and not in the place where the object was used, so it often is with fossil hominids. The one thing we can be sure of is that the fossil record tells us where an animal died, not where it lived, and still less *how* it lived. The fossil record is distorted, and we have to bear this in mind if we are not to be led up interpretative blind alleys.

One trick that is helping prehistorians steer clear of these unproductive paths is ethno-archeology, the sort of exercise that, for instance, Diane Gifford has been doing on the contemporary camps of the Dassanetch people. Perhaps the most striking lesson that comes from this work is the possibly distorted view we get of the location of living sites. The Dassanetch use two kinds of camps: temporary camps that they occupy for just a few days and which may consist simply of a hearth with no other major structures; and more permanent pastoralist settlements, known in East Africa as manyattas, which are encampments surrounded by thorn-bush fences. Because, naturally enough, these people depend on a reliable water supply, most of their camps are within striking distance of the Turkana lakeside or a large sand river where wells can be dug. These people are well aware of the effects of seasonal rains and of fluctuations in the level of the lake waters. They settle their camps accordingly: manyattas are on raised ground where they might avoid devastation by flooding, whereas the temporary camps are perched much more precariously near the water's edge.

From the occupants' point of view, a camp covered in water is more than a little inconvenient. But for the archeologist it is virtually the only hope that a campsite will be fully preserved. The fine silt that is deposited in steadily-rising lake waters, or by flow in the seasonal streams, is extremely favorable for preserving both the bones and stones left on the abandoned camp. Camps located close by the lake and along the banks of streams are therefore more likely to be preserved than those built safely on higher ground. For the Dassanetch this means that an archeological survey will uncover mainly the temporary living sites whereas most of the more permanent settlements will have disappeared.

Could this be the case with our hominid camps? We have to admit that most of the occupation sites are in positions that are vulnerable to flood: lake margin or the bed of seasonally-active streams. Yes, these must be temporary sites. But it is possible that our hominids never occupied anything other than temporary camps. This is not unlikely, given their probable life-style which we shall describe shortly. Even if a succession of temporary camps was the normal existence for the Turkana hominids, however, the margin of the major lake may not have been the only place where they pursued their economies. To illustrate this warning we can look to the !Kung of Botswana, a hunting and gathering people who subsist mainly on plant foods but who hunt animals – mostly small – as a minor part of their diet. They too live by permanent lakes, but only in the dry season. When the rains come they disperse in bands and obtain water from temporary water holes. In the case of the !Kung, their dry-season camps are much more likely to be preserved than those they occupy in the wet season, even though all the camps are relatively temporary.

Just to complicate matters further, neighbors of the !Kung, the G/wi people, do just the opposite. They move their bands to waterholes in the wet season and disperse during the dry season. The G/wi are rather unusual, however, in that they manage to subsist for ten months of the year with no standing water what-

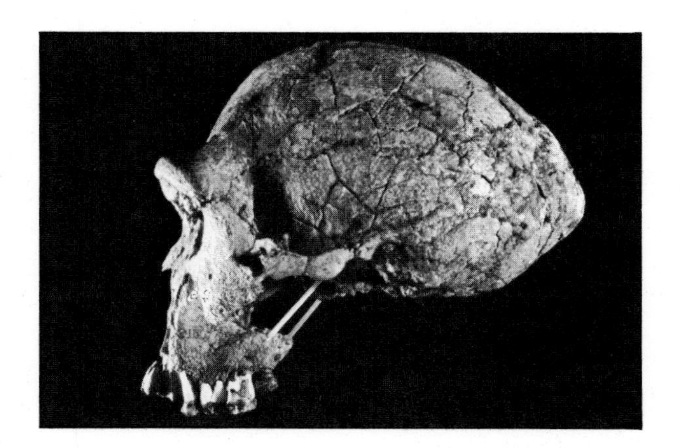

This nearly complete skull of Homo erectus *found at East Turkana in 1975 is particularly interesting because its cranium is very similar to that of Peking Man's. Dated at about 1.6 million years, it shows the rapid pace of human evolution at this time.*

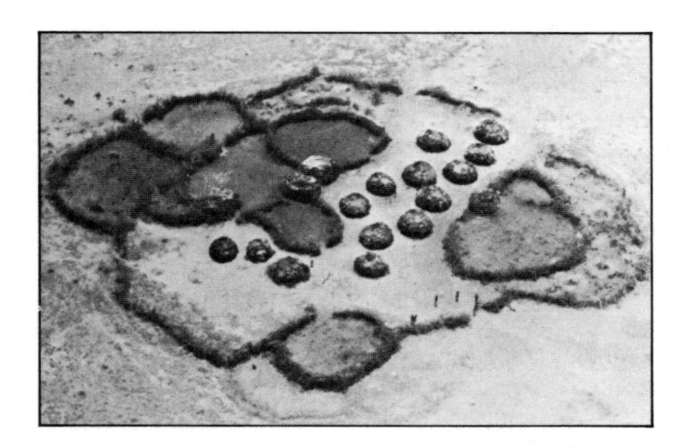

An aerial view of a Dassanetch manyatta. The animal pens and outer fences are made from thorn bushes. Small one-family houses, which are completely portable, occupy the central area.

Top: Richard and Meave Leakey, and Bernard Wood, near Lake Turkana, excavating a skull of Australopithecus cf. africanus. *The preliminary reconstruction of it is shown above and, right, a comparison between the skeletons of* Australopithecus africanus, Australopithecus boisei, *and* Homo sapiens sapiens. *Although both* africanus *and* boisei *walked upright they may have been less well adapted to do so than early* Homo.

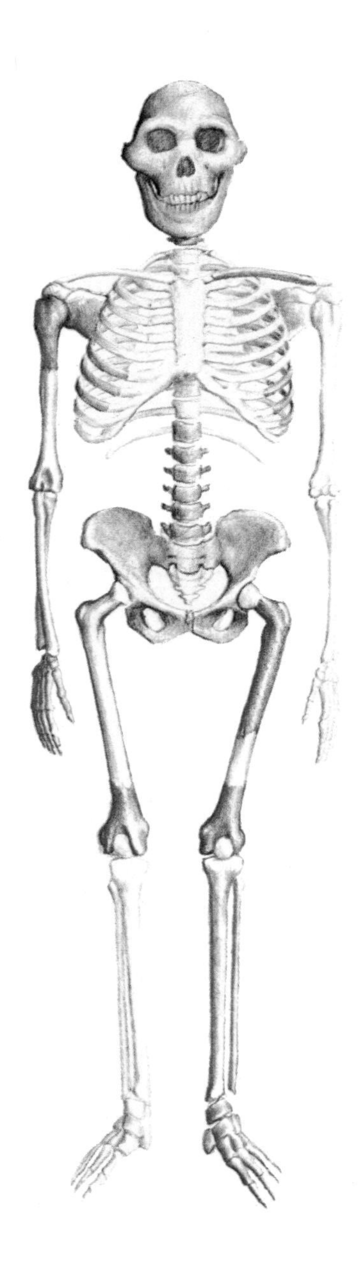

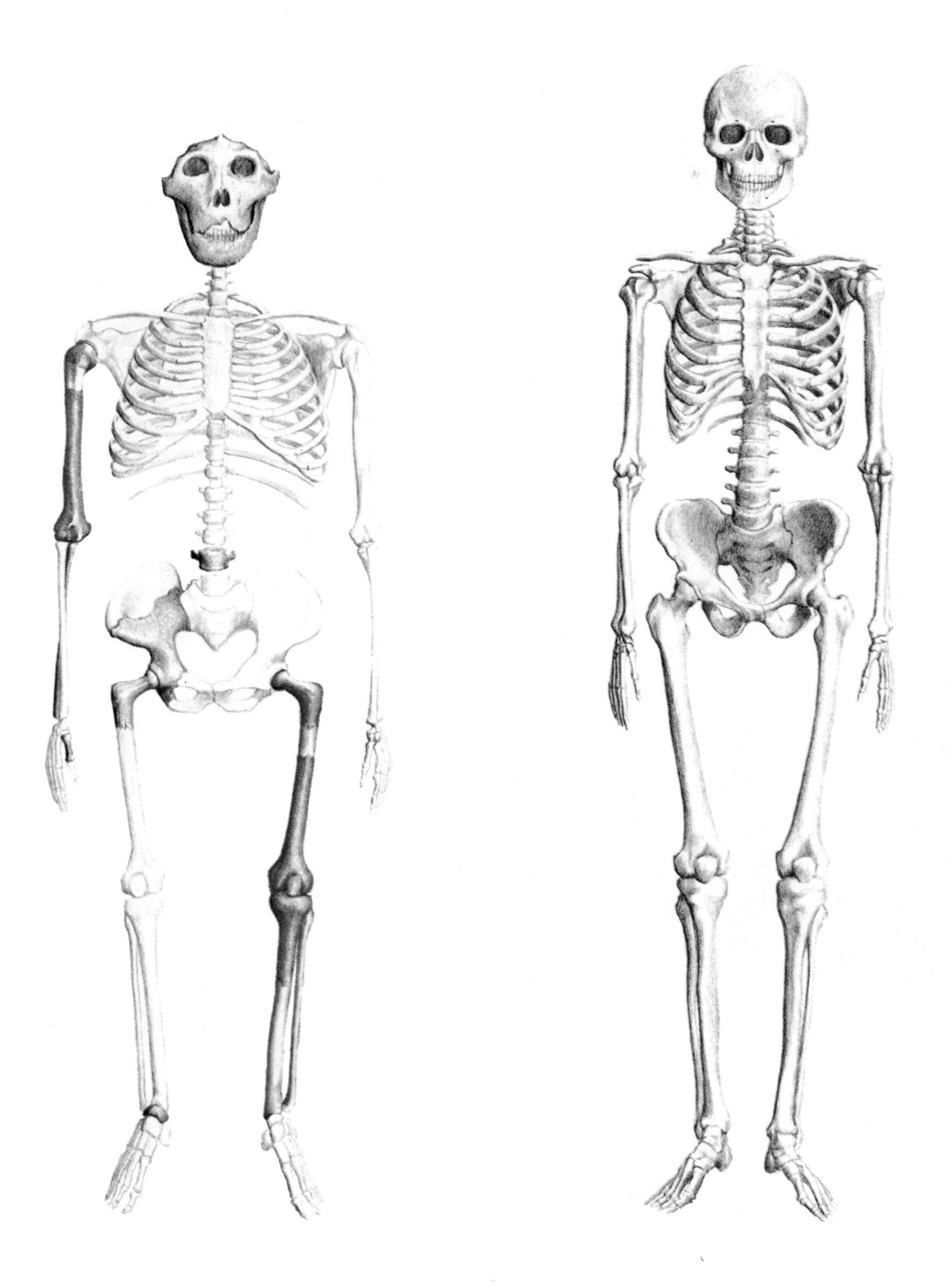

Taking all their personal belongings, these !Kung people are moving camp to another waterhole.

G/wi woman digging for succulent roots in the Kalahari Desert – Khutse Pan, Botswana.

soever: they obtain their liquid intake by eating succulent plants. The G/wi experience emphasizes the possible complexity of life in people who live in temporary camps. Overall, there is a definite possibility that the lakeside camps we find from two million years ago represent dry-season occupations, like the !Kung's, and that during the wet season bands of hominids strayed from the lake leaving campsites that would be very difficult to locate, and anyway probably were largely destroyed. Only a careful statistical analysis of the bones and artifacts on the occupation sites we do find will tell us if we are being misled.

Another feature of hominid camp life that ethnoarcheology may give us an insight into, is the meaning of the distribution of tools and debris. The emphasis here, however, must be on debris as no contemporary peoples use stone tools exclusively. One sees that in camps that are used for just a very short time there is no special effort at housekeeping. Small bones and other food refuse do not build up sufficiently to create a nuisance. But in home bases that are occupied for longer periods, the litter problem does impinge on camplife, and then systematic efforts are usually made to clear away the debris, generally to a nearby place. These sort of patterns occur in the Dassanetch, as Diane Gifford has observed, and there are persuasive parallels at Olduvai Gorge.

One of the most remarkable discoveries at the Gorge is a two-million-year old rough circle of stones, some of which are piled one on top of another, forming what seems to be the oldest man-made structure yet known. Very probably the hominids who occupied this site stacked branches in a circle, using the stones to keep them firm. Within this shelter, if indeed that is what it is, there were only thirty pieces of bone and stone fragments, a very low density compared with many occupation sites. Either the occupants cleared out their refuse, or they manufactured their tools and prepared food elsewhere.

At another Olduvai site, the occupation floor associated with the first *Australopithecus boisei*, there is a dense concentration of bone and stone fragments, but this borders on a nearly empty area. It is a fair guess that the empty area represents a place where the camp dwellers sat and ate, and that the concentration of debris was their rubbish tip. A wind break or some such structure may have separated the two areas. Incidentally, just because the cranium of the robust australopithecine was found there, does not mean that the occupants were also *Australopithecus boisei*. Indeed, the chances are that they were not; but of this more later.

If we now took another trip in our time machine and aim for a point east of Lake Turkana about two and a half million years ago, what would we see? The scenery would not have been so very different from what we see today. The lake was higher and, in the same way as today, it was fed by seasonal streams and rivers which are bordered by large trees and bushes, depending on the amount of ground water there was available. The strip near the lake is green with grass of unusual lushness for an otherwise mainly arid area. Gazelles, waterbucks, pigs, and perhaps the occasional giraffe come to drink from the lake, but they would need to be wary of the crocodiles lurking in the shallows. Hippos and catfish are also visible from time to time. In the far north smoke rises silently into a blue sky, a reminder of a recent eruption from one of the volcanoes in the Ethiopian mountains. Soon the ash that plumed into the skies and cascaded down the volcanic cones will be carried down the rivers and streams to be deposited in the ever-thickening layers alongside Lake Turkana.

If we had been thoughtful and had remembered to take a Land Rover with us we could drive northwards from the sweep of Alia Bay, past the Koobi Fora spit where the East Turkana Research Group now has its camp, past the Karari escarpment on our right, and into the low hills of Ileret, a journey of some eighty very bumpy kilometers. In the day's outing we would, if we were lucky, have spotted two or perhaps three groups of hominids, but they would have taken to their heels very swiftly at our noisy approach. But if eventually we learned to approach cautiously, with the benefit of strong field glasses we would, after a dozen or so forays, have determined that there were indeed four different types of hominids living by the lake margin and surrounding hills, some of which are more common than others. We would also have noticed that, not only do the hominids differ in stature between the separate types, but their behavior patterns vary too; however, one of the hominid types is strikingly distinct from the other three. This behavioral diversity should not surprise us because animals that are basically the same type, as the hominids certainly were (they all walked more or less upright for instance), must operate separate subsistence economies in order to coexist.

The type we see most frequently is a robust individual, with massive jaws and standing around five feet tall. The shape of the head betrays the presence of big muscles which move the massive jaw, and just as it is in modern gorillas, this is much more pronounced in the males than in the females, which overall are smaller anyway. As we watch a troop of twenty or so individuals moving slowly, spread out among the bushes and grasses, we can see the reason for the powerful jaws. Each individual, feeding by itself, is searching for roots, grass seeds and other plant materials, all of which are tough and need pulverising between the massive molar teeth set in those sturdy jaws. They pull roots from the ground, mostly using their hands, but occasionally employing a stick to loosen a particularly troublesome but promising morsel.

The group moves along in an apparently coordinated, but unhurried way, each individual searching for food and chewing as it goes. In spite of their solitary

Overleaf: An impression of life in a campsite occupied by Homo habilis.

feeding habits, they are clearly social animals; the interactions and alliances confirm that. And as the sun begins to set behind the western wall of the Rift Valley, the troop tracks back to the rock face or clump of trees where they slept the previous night, a resting place that gives some protection from the big carnivores that seek out their prey as night falls or just before the sun comes up the next morning.

This is *Australopithecus boisei*, an animal not unlike a baboon in its feeding habits, but lacking that modern monkey's catholic taste. In fact, *boisei* is much more like a wart hog in its tastes, if not in appearance! This hominid has obviously become adapted to a tight ecological niche, and its massive jaws and huge millstone molars are the result.

While we are out watching troops of *boisei* we would see almost as frequently bands of very similar hominids, though slighter in build and somewhat more mobile. This is *Australopithecus africanus*, a creature which is occupying much the same niche that modern baboons now thrive in. Like their *boisei* cousin, *africanus* gathers in social groups, though they contain more individuals. Roots and seeds are also on *africanus*'s menu, but so are berries, nuts, grubs, beetles, small lizards, birds' eggs, and even the occasional baby gazelle that is separated from its mother. They use their hands with some agility, and occasionally pick up a sharp-edged stone to scrape into the bark of a tree in search of succulent grubs. Occasionally, too, one of the troop cracks a smooth pebble on a rock, a flake flying off to leave a useful sharp edge, but there is no sign of systematic tool-making. They also use rough stones to pound through the tough shells of protein-rich nuts. These creatures are tool users, but not adept tool-makers. The hallmark of their lifestyle is opportunism.

As darkness begins to descend on the lakeside, heralded perhaps by one of the most spectacular sunsets you could hope to see, with the sky streaked red from horizon to horizon, the troops of *Australopithecus africanus* follow the example of their stockier cousins, and head for a safe resting place. The same trees that offer comparative safety during darkness are also climbed with impressive agility at the threat of danger, caution rather than retaliatory aggression playing the dominant role in their reaction to predatory threats.

Probably the rarest hominid sight by the lake at this time in history are the survivors of the original *Ramapithecus* population. Smaller than *Australopithecus africanus*, their lives are very similar to their descendants, except that they are even more agile in the trees, having feet better suited for climbing. But the overlap in ecological niches between these late members of *Ramapithecus* and the bigger *Australopithecus africanus* is too great, and before long the more ancient hominids succumb and sink into extinction.

Although it is relatively easy to see why *Ramapithecus* should become extinct, it is less so for the two australopithecines. According to the fossil record they both disappear at around a million years ago. And the emergence then of 'modern' baboons may well have something to do with the eclipse of *africanus*. Perhaps this hominid was squeezed out of its niche by a pincer movement: on one side were the baboons who were steadily establishing themselves on the savannas, and on the other were the human ancestors, *Homo erectus*, who were fast becoming enormously successful and more and more populous. There is no evidence to support the notion that the australopithecines were propelled into extinction through appearing regularly on the menu of *Homo erectus*.

It has been suggested that *Australopithecus africanus* was the first of the two australopithecines to disappear. The bigger species may have become extinct a million years ago, but as it was less in competition with either baboons or *Homo erectus* than was its slighter cousin, *boisei* may well have survived substantially longer. As yet there are simply too few good archeological sites dated between one million and three hundred thousand years ago – we cannot be certain that small isolated populations of *boisei* did not survive until as recently as the latter date. It would be remarkable if they had, but it does linger as a tantalizing possibility.

But we have not yet finished our description of the Turkana lakeside two and a half million years ago. The last of the four types of hominids, though a relatively rare sight compared with the australopithecines, is by far the most intriguing. They are very similar physically to the two australopithecines: in size they compare roughly with *boisei*, but they lack the massive jaws and associated bony crest on their skull; their heads appear to be slightly bigger than the australopithecines, but not dramatically so; they appear to walk with an easier stride, but again, the difference is not dramatic. But what is strikingly distinct is their behavior.

We do not see troops of this hominid making their way gradually through the bushes, each individual eating as it goes, males, females and infants together. The groups we see are smaller, and instead of eating the nuts, fruits and grubs as they pick them, they gather the food and carry it away, using a crude container woven from long leaves. Now we see two more, each carrying the leg of a pig which they sliced off a fresh carcass they came across in their search for nuts and fruit. The two individuals carrying the scavenged meat follow the path taken by their fruit-gathering companions, and eventually arrive at a tree-shaded camp in a dry stream bed.

This is a camp in the real sense: the occupants have been here for several days already; they sleep there each night; and it will be some time before they move on. Meanwhile, members of the band go off in twos and threes, returning later in the day with a rich mixture of foods: mainly plant foods, but also some small rodents, and occasionally bigger pieces of meat, such as scavenged pig. Most of the meat comes in this way, and there is enough carrion around to make it worthwhile actively searching for it. Giraffe, porcupine, and several species of gazelle find their way onto the menu by this route. One young gazelle is actually caught alive after a brief chase. But this is unusual.

Like the other hominids, these camp dwellers are also social, but to a noticeably higher degree: there are more intricate interactions between individuals, and not just between kin; they seem able to communicate with each other by a combination of complex sounds (almost sophisticated enough to be considered as language), and gestures. The focus of the camp is the sharing of food. Although the division is by no means exclusive, it is the males who generally spend some of their time scavenging for meat prizes, and looking out for the occasional opportunity to hunt. The males certainly collect plant foods too, and the females do not pass by the chance of meat if they find it. But the active search is, in the main, a male preoccupation. In any case, if any individual tried to live by eating only meat, he would probably starve: the day-to-day menu is mostly plant foods, with the occasional bonus of antelope, pig, or whatever else is chanced upon.

After perhaps ten days the group moves on, leaving behind the discarded bones and stone implements they made to fashion digging sticks, slice meat, and scrape skins. As they move off a fig leaf drifts slowly down from a nearby tree, and lands softly in the abandoned camp. The site is never occupied again. Instead, gently, very gently, it is covered with fine sand and silt, to be entombed for two and a half million years, when, first by the ravages of weather erosion, and then by the archeologists' tools, it is exposed once more, giving us clues of human life long ago.

All of this is, of course, a complete fairytale, a fabric of more or less inspired guesses. The truth is that no one knows exactly how the hominids lived. But our guesses are based firmly on the few clues that we do have, and they slot into a reasonable biological scenario.

The core of the behavioral differences between the australopithecines and the ancestral humans, *Homo*, is simple but crucial: *Homo* established home bases and they shared their food. Our ancestors switched from being opportunistic food eaters to being systematic food gatherers. The more frequent addition of meat to the menu was a valuable source of high-quality protein. But the exchange of specific foods between individuals, something that no other primate does to any important degree, had deep behavioral and social implications for the emergence of humanness, a topic discussed more fully in Chapter 7.

Our ancestors at this time, therefore, were gatherers, both of plant foods and of meat. Although the practice of hunting as opposed to scavenging undoubtedly increased as time passed, the chase and the kill have achieved an undeserved status in the popular conception of human ancestry. Our ingenious ancestors very probably learned many hunting techniques, and on occasion employed them to bring down quite large animals. But it was by no means the mainstay of the hominids' economy. Some people suggest that the social and psychological aspects of organized hunting were the prime movers in human evolution. Although these factors certainly played an important part in molding humanity, we see them as secondary to the real prime mover of our emergence from a more primitive hominid stock; and that is the practice of sharing food within an organized social group. It was the biological legacy of a food-sharing economy and social organization that equipped our ancestors some one and a half million years ago for the journey into Asia and Europe.

Africa was the cradle of mankind. By the time *Homo erectus* emerged our ancestors were equipped to explore the rest of the world.

6 From Africa to Agriculture

The bands of our ancestors, *Homo erectus*, who around a million years ago made their way across the narrow strip of arid land that joins the continent of Africa to Asia were in the vanguard of mankind's ultimate domination of the earth. Contrary to the once popular belief, the gradual occupation of Asia and then Europe must not be seen as a drive by an avaricious people hungry to take over new lands. Nor did it represent the abandonment of Africa, which would then have to wait several hundred thousand years before being repopulated by truly modern humans born in more 'civilized' parts of the globe.

No, the spread of *Homo erectus* into the northern continents was an inevitable consequence of evolutionary momentum, of a social and behavioral organization based on the unique human trait of food-sharing. These people were now equipped mentally and technologically to deal with any challenge their world might confront them with. And who can deny that by that time there was already kindled in the human mind a spirit of adventure, a real curiosity about the world around them? This newly emerging element of humanity must have combined with the constant search for new sources of food to produce a gentle, unhurried exploration of new lands. Certainly there was no steady climatic change to draw ancestral populations northwards. Indeed, the world was then experiencing one of the most turbulent periods of its climatic history, marked by frequent advances and withdrawals of the great northern ice sheet. Bands of *Homo erectus* undoubtedly took advantage of every favorable opportunity offered by the environment, but the fundamental reason for their success in globe-trotting was within themselves, not outside.

While trying to single out the unique trait in the minds of humans some million or so years ago, we must be careful not to exaggerate. Unquestionably, the mind of *Homo erectus* was special. After all, evolutionary forces were propelling it rapidly towards that remarkable structure that is in the heads of all of us: the brain of *Homo sapiens sapiens*. But, although our ancestors, observing the seasonal arrivals and departures of great flocks of migrating birds, must have wondered where they came from and where they were going to, and speculated on what might lie on the other side of distant hills, they themselves almost certainly did not embark on major migratory expeditions. They may well have scaled the distant hills to satisfy their curiosity, and, perhaps liking what they saw, settled their new camp there. Their migration as such, however, was a slow business, populations moving no more than a few tens of miles in a generation – though it is salutary to remember that even at the modest pace of ten miles a generation the journey from Nairobi to Peking, for example, would have taken less than fif-

Previous pages: Cro-Magnon Man was an excellent artist and many paintings have been found in the caves he inhabited – particularly in France and Spain. This example of a bison is from the Altamira caves, near Santander, Spain.

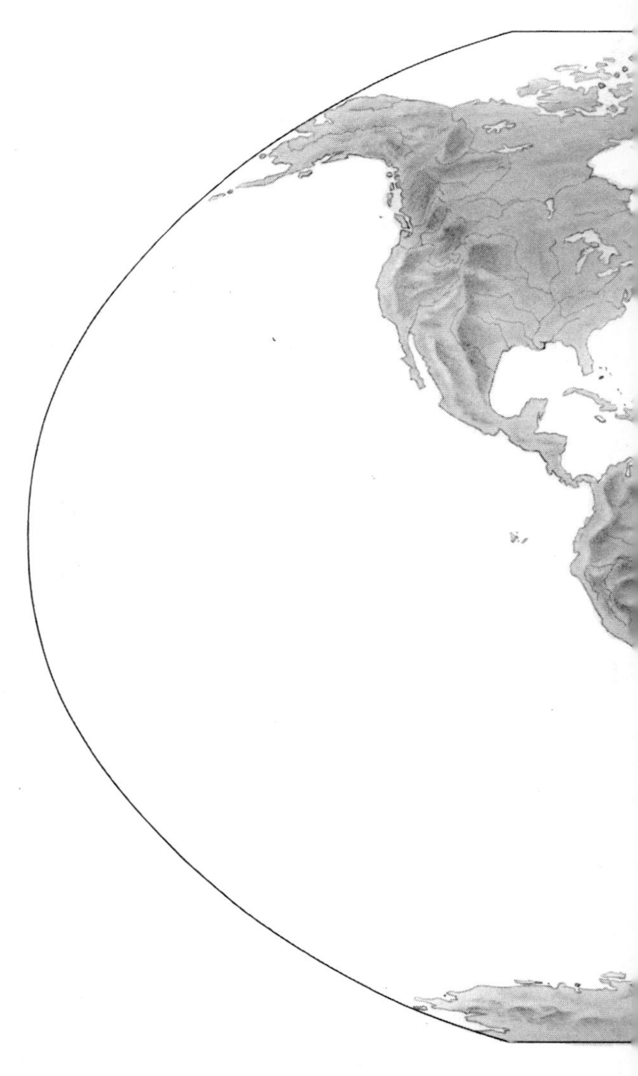

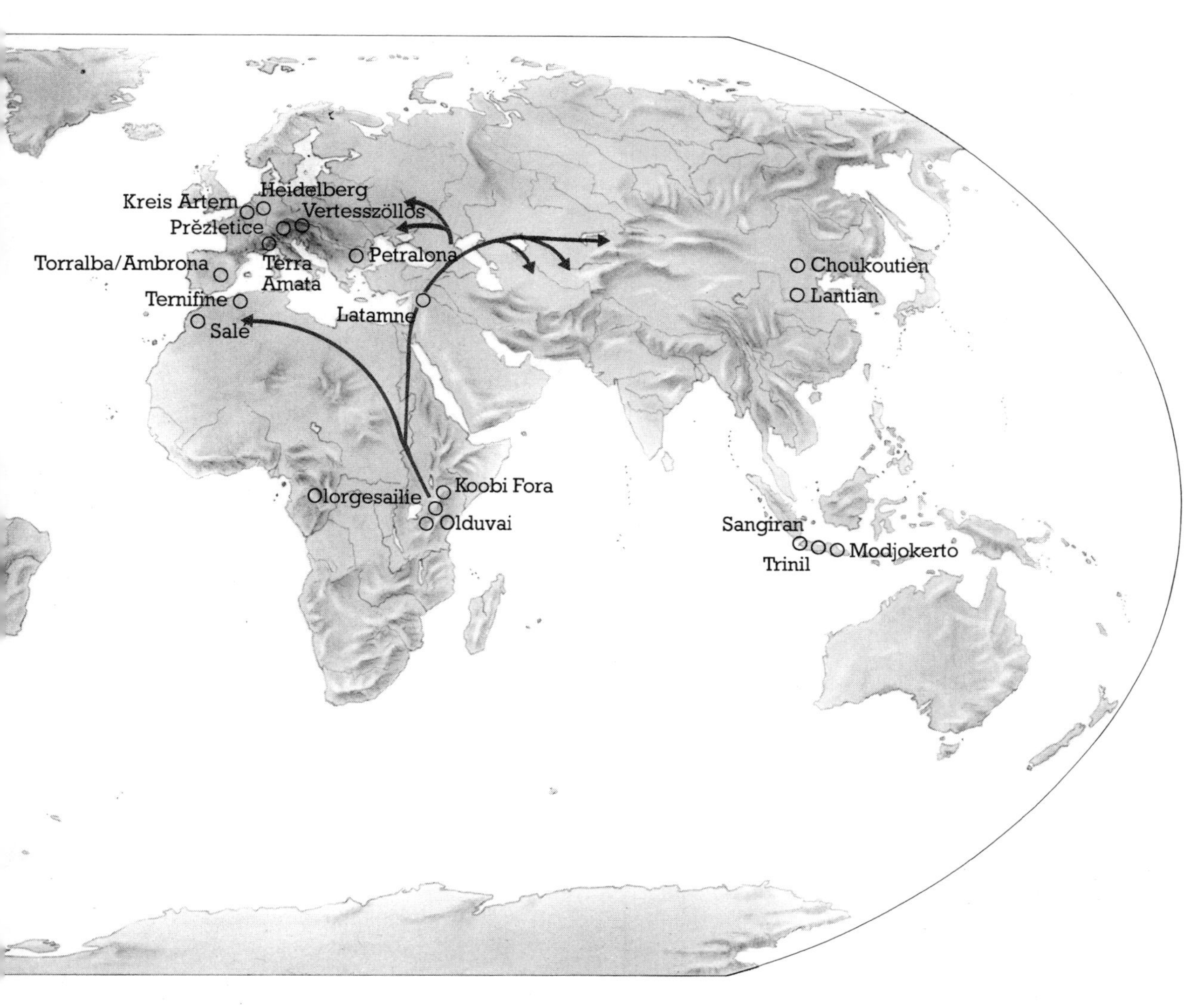

Evidence of Homo erectus *has been found at sites over a very wide area of the Old World. The most important localities are given this map, which also shows the routes he probably took as he expanded from his tropical world into temperate regions.*

teen thousand years – in evolutionary terms a very rapid journey indeed.

Between the time that the *Homo* line was born alongside its australopithecine cousins in Africa, five or six million years ago, and the point at which, for whatever reason, bands of *Homo erectus* began spreading into the rest of the world, human prehistory has been a matter of evolutionary forces that have shaped and molded the physical structure of prehuman creatures. From a million and a half years onwards, and particularly where the past three hundred thousand years are concerned, the center of interest changes: it is now the head alone that commands attention. If, by some magic, a *Homo erectus* individual attended a masked party – Halloween for instance – in twentieth-century London or New York, his stance and general appearance would have occasioned no special comment; a little on the short side perhaps, but nothing out

of the ordinary. But what a shock the other guests would have had as midnight approached, and the time came to discard masks! Our atavistic guest would have a strangely flattened skull, prominent brow ridges, and a protruding jaw. And if anyone had cared to look, his molar teeth would have appeared as much bigger than a modern dentist would be accustomed to see.

The raising of the cranium to a domelike shape, thus accommodating a bigger brain, the reduction in the size of the molar teeth, and the protrusion of the jaw are basically what this last major phase of human evolution is all about: such is the transition from *Homo erectus*, through the basic *Homo sapiens*, and finally to fully modern humans, *Homo sapiens sapiens*. Brain size in *erectus* populations through time ranges from around 775 cc to 1300 cc. This compares with the variation in modern humans of from 1000 cc to 2000 cc, with an average of around 1400 cc. Some *Homo erectus* individuals therefore had brains bigger than some people living today! The total *size* of the brain is not, however, absolutely important in terms of intelligence: people with bigger heads are not necessarily brighter or more intellectually resourceful than people with smaller ones. A steady trend in human evolution has indeed been an increase in the size of the brain; but what is equally important – though totally invisible in the fossil record – is the internal reorganization that made possible new and better nerve networks and brain centers. Here is the key to the final stages of our evolution.

Applying a rough time scale, we can say that the step from *erectus* to *sapiens* occurred around half a million years ago, and the refinement to *sapiens sapiens* perhaps fifty thousand years ago. These transitions must have occurred not just once, through some kind of omniscient predestination, but many times and in many places, as the built-in evolutionary momentum propelled *erectus* towards *sapiens* in a biologically unstoppable way. Certainly, isolated populations of *Homo erectus* must have been left behind in the process of evolutionary advancement. And occasionally, too, some *sapiens* populations would seem to have veered along ill-fated evolutionary routes from which they could not retreat; the stocky ice-adapted Neanderthalers (more properly called *Homo sapiens neanderthalensis*) of Western Europe are an example.

Overall, though, our ancestors in Europe, Asia, and

Reconstructions of Gigantopithecus *(left) and* Homo erectus *(right). The evidence from the finds of the former suggest that it was a large ape-like genus that was adapted to ground feeding in the open country of Central Asia and Northern India.*

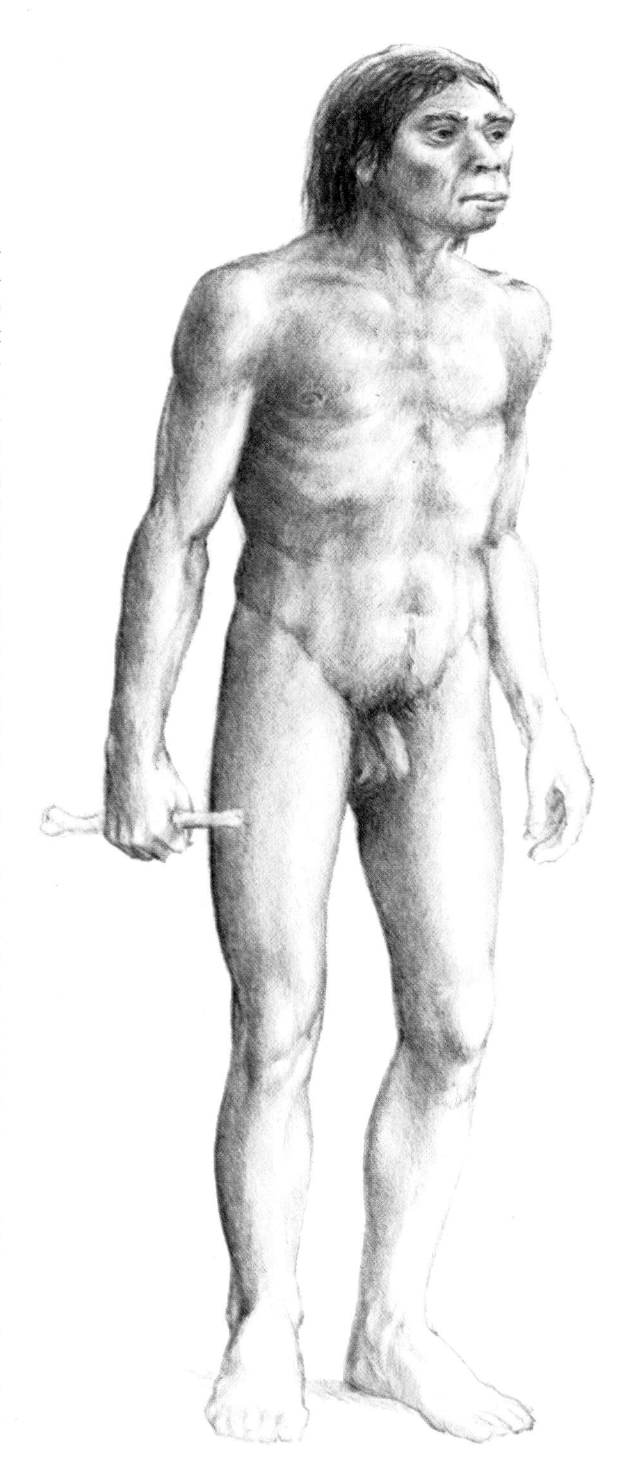

Africa represented a pool of human genes which, assembled and reassembled into better and better combinations, would eventually produce mankind as we know it today. And genes were not the only things that were being exchanged between neighboring populations: stone tool cultures came and went in waves, quite as fashions do today. The pace of human advancement, both biological and cultural, during these last stages of human evolution is dizzying compared with the relatively sedate progress of the previous three or four million years. And there can be no doubt that the sharpest and most dramatic shift of gear in our ancestors' progress along the path of human evolution was the invention of agriculture ten thousand years ago. That shift from an essentially mobile hunting and gathering existence to an essentially sedentary agricultural economy, shattered a way of life that had first emerged at least three million years earlier, and was responsible for creating the basics of humanity in the way they are. The invention of agriculture was, without exaggeration, the most significant event in the history of mankind.

When *Homo erectus* set out from Africa on the path that would lead eventually to the agricultural revolution, it was to enter a world devoid of other advanced hominids – or so it appears from the fossil evidence so far. Why? We have no good answer to the question of why *Ramapithecus* gave rise to *Homo* and the australopithecines in Africa, beyond saying that some kind of ecological change offered new niches to be filled by hominid-like creatures, so we cannot be firm about why the same basic stock did not give issue to the same descendants in other parts of the world. We can, of course, say that the ecological opportunities offered in Africa did not arise elsewhere, and that this may have something to do with the balance of forest and savanna: the shrinking forests in Africa, opening up greater areas of open woodland and grassy savanna, may have been just the right conditions to nurture a newly emerging hominid family; and possibly the same kind of shift did not occur to the same extent in Asia and Europe, continents which were also populated by *Ramapithecus*.

Another guess is that the equivalent niche in these continents was occupied by a creature not itself a hominid, but with habits similar enough to impede the emergence of a new hominid type. An obvious candidate would be *Gigantopithecus*, the large ground-living ape, about the size of a modern gorilla, that lived in Asia from about nine million until perhaps no more than one million years ago. As an opportunistic omnivorous ape, living much as the gelada baboons do in the highland plateaux of Ethiopia, *Gigantopithecus* might well have been enough to discourage the intrusion of the much smaller early hominid into an overlapping ecological niche. But evidence to support this notion is at best minimal.

One thing we can be certain of, however, is that even before members of the *Homo* lineage began moving into Asia and Europe, its evolutionary progress had begun to accelerate. The recent discoveries of remarkably developed yet very early specimens of *Homo erectus* in the fossil-rich deposits east of Lake Turkana tell us that. Just a million years separate the individual we know simply as 1470 from the *Homo erectus* who turned up in 1975. The former was most definitely *Homo*, but lacked the more advanced features of *Homo erectus*, such as a bigger brain, rounder cranium, flatter face, and prominent ridges above the eyes. There is, incidentally, a good deal of regional variation between the different populations of *Homo erectus*: for instance, the individual found at Olduvai in 1960 had unusually large brow ridges. Such variations are just what you would expect in a species evolving rapidly in different geographical locations. And this phenomenon was particularly marked in the last million years of human evolution, with Asia, Europe, and Africa presenting something like a mosaic of more or less advanced forms of humanity. But it was a mosaic in motion, with separate populations interacting both culturally and biologically, interchanging tool technologies and genes.

At any rate, one oddly simple but basic fact has figured enormously in the passage from the ancient *Ramapithecus* stock straight up to the emergence of modern humans: the ability to carry things. The concept itself is unarguably simple, but it is fundamental because of the unique degree of independence from the environment that it confers on humanity. We are not talking just about the ability to transport food, but of three other commodities too, each of which at different stages in the journey of human evolution helped propel our ancestors along the path to mankind.

The ability to carry food was part of the behavioral package that transformed a forest dwelling ape into an upright-walking hominid. Transport of food was also essential to the mixed economy of hunting and gathering centered on a home base. Next comes water. Animals are highly dependent on water, and hominids are certainly more dependent than most. But either by using large egg shells, gourds, or simply taking along a succulent melon, they could greatly increase the range of their hunting and gathering. Transportable water may also have been an important factor in the step-by-step migration across the arid land strip that joins Africa to Asia. Third among transportable commodities is fire, an almost magical phenomenon: its sensual fascination is experienced by people throughout the world, but it would have been particularly important as our ancestors wrestled with the colder climates of northern Europe. Finally comes the ability to transport experience itself – from individual to individual and from generation to generation. The utensil in which it is transported is, of course, language – a development in which *Homo* had already departed substantially from primitive ape-like vocalizations as much as two million years ago. In the transition from *Homo erectus* to *Homo sapiens sapiens*, language was surely central in knitting together the social and cultural structures of the mingling populations.

The independence made possible by these four transportable commodities is the key to the remarkable migration of our ancestors from tropical Africa to the ice-fringed continents of Europe and Asia.

The discovery of fossil remains of what is now called Neanderthal Man has been described briefly in Chapter 2 (see p.32). Possibly because he was the first obviously archaic human to be unearthed, and because so many fossils have since been found (there are now bits and pieces of more than one hundred individuals), Neanderthal Man has become fixed in the minds of many people as the archetypal human ancestor: a low brow; a thrusting face, but with a receding jaw; fearsome beetle brows; and a stooped, lumbering gait in which a stocky muscular body was dragged about with seemingly malevolent intent. Misconceptions about the Neanderthalers' posture came mainly from the relatively complete but severely contorted remains of an old arthritic individual who died at what is now known as La Chapelle-aux-Saints in southern France. The notion of malevolence came from nowhere but a hostile imagination.

We can now be sure that the Neanderthalers led a complex, thoughtful, and sensitive existence, surviving somehow in the extremely harsh conditions of an ice-gripped Europe. But we can also be sure that these people were not direct ancestors to modern humans. The genetic pool that was eventually to yield modern man also produced the Neanderthal people. Specialists in coping with the cold, they finally succumbed to the evolutionary trap of becoming *too* specialized. When the glaciers began to recede at the end of the last ice age these people were too rooted in their ways, both biologically and behaviorally, to adapt to the warmer times that lay ahead. That warmer climate was to be effectively exploited by the more general human stock that eventually emerged as *Homo sapiens sapiens*. The discovery of an early form, the Cro-Magnon, has also been described in Chapter 2 (p.32).

By the time the Neanderthal populations slid into eclipse around thirty thousand years ago, truly modern humans had been firmly established for at least twenty thousand years. But there is no convincing evidence to suggest that waves of modern man swept through Neanderthal territory, raping, pillaging, and murdering all who stood in their way. Pockets of Neanderthalers, biologically far along their evolutionary blind alley, would have remained separate from the newcomers until they died out through economic competition. But others who were genetically less distant from the evolving *sapiens* populations might have been absorbed by interbreeding. Although some blood may have been spilled in the encounter, there is no good reason to conceive of the whole episode as a bloody period in human prehistory.

Neanderthal-like people were not confined to the northern-most areas of Europe: their remains have been found in France, Spain, Italy, Yugoslavia, Iraq, China, Java, Zambia, and Israel, among other places. In many of these areas there are rich collections of tools, and from the variety and subtlety of the implements their abilities are obvious: cloth-making and delicate carving must have been well within their means, and the patterned differences between different tool assemblies tell us that there were at least four coexisting cultures, four tribes perhaps. In the midst of all this, one relic strikes a particular chord of humanity. It comes from the Shanidar Cave in the Zagros Mountain highlands of Iraq, where on a June day some sixty thousand years ago a man was buried in unusual circumstances.

The humidity of the cave was far from favorable for preserving the bones of the dead man; but pollen grains survive very well under these circumstances, and researchers at the Musée de l'Homme in Paris who examined the soil around the Shanidar Man discovered that buried along with him were several different species of flowers. From the orderly distribution of the grains around the fossil remains, there is no question that the flowers were arranged deliberately and did not simply topple into the grave as the body was being covered. It would appear that the man's family, friends, and perhaps members of his tribe had gone into the fields and brought back bunches of yarrow, cornflowers, St Barnaby's thistle, groundsel, grape hyacinths, woody horsetail, and a kind of mallow. The branches of the woody horsetail are particularly suitable for weaving a rough bedding on which the body appears to have been laid; and the white, yellow, red, blue, and purple flowers of the other plants must have added greatly to a poignant scene.

The fact of deliberate burial is interesting enough, for it betrays a keen self-awareness and a concern for the human spirit. And to have decorated a corpse with flowers adds enormous significance. But perhaps most intriguingly of all, it turns out that of the various species used in the Shanidar burial, several have until recently been used in local herbal medicine! That the Shanidar people were aware of at least some of the medicinal properties of these flowers is not unlikely. As they became more and more independent of the demands of the environment, early members of the human family must also have had an intimate knowledge of what Nature offers. This, indeed, is the key to success among hunting and gathering communities.

The Shanidar burial therefore gives us an insight into the spirit of Neanderthal Man, and also a hint that his culture had gone beyond the shaping and using of stone tools: from a close relationship with nature he also learned some medicine. The findings at Shanidar are at least as important in telling us something about the minds of our true ancestors as they are direct evidence of the ill-fated Neanderthalers. Although as yet there are no signs of ritual as subtle as the flower burial for our true ancestors, we can be sure that their culture was no less developed.

Overleaf, an artist's impression of the Shanidar cave burial.

The burial finds at Shanidar in Iraq are remarkable because they give us an insight into a cultural aspect of Neanderthal Man. The photograph above shows the site of the cave that was examined by Ralph S. Solecki and his team. Left, a view of the flower burial as it was excavated. Below, a skull as it appeared when cleared from soil.

The elaboration of ritual and culture are, of course, crucial signals of emerging humanity in our primitive ancestors. A tantalizing clue to the early expression of that culture comes from Terra Amata, a hillside site at Nice, in southern France, where some four hundred thousand years ago a band of hominids made camp. In the usual pattern of archeological discovery, workmen turned up some stone tools while clearing the area for a block of luxury apartments. The site was extensively excavated by Henry de Lumley of the University of Aix-Marseille with the help of scores of volunteers, who eventually uncovered the remains of what must have been substantial living quarters. By digging down through the living floor they found evidence that the site had been occupied over and over, probably at yearly intervals, and possibly by the same people.

The occupants of the camp had built their shelters by driving wooden posts into the ground, using rocks to give them stability, and then covering the whole, possibly with animal skins, branches, or a combination of the two. From the size of the camp it would seem that

A reconstruction of a Terra Amata shelter.

Further evidence of early man's habits were found in Nice, on the French Riviera when Henry de Lumley was examining a site (left) being bulldozed in preparation for a block of apartments, near an alley called Terra Amata. Among the many thousands of objects found was evidence of the type of shelters erected by the occupants of the campsite. The shelters' construction can be surmised from the bracing stones that were found (above left) and the imprints of stakes.

there had been about twenty-five people in the band; when they abandoned the site they left bones of deer, elephant, wild boar, and other animals. Along with these there must have been a scattering of vegetable matter too, but as usual this would have vanished from the fossil record. These people undoubtedly were hunter-gatherers, but as there are no human bones near the camp we can simply make the reasonable guess that they were *Homo erectus*, and possibly a borderline type on the way to becoming *Homo sapiens*.

That these people, whoever they were, had slept in their huts, huddling close to the hearth for warmth and comfort, we know because the area around the hearth is clear of debris; bones and stones underfoot would have interfered with sleeping. They also left behind what was probably a crude wooden bowl, and near to it there were a few lumps of ochre, the natural red/orange pigment, that had been sharpened to points so that they could be used for drawing. What it was that these camp-dwellers drew, and where they drew it, we shall never know. They may have been decorating their bodies for some kind of springtime festival (that it was a springtime camp, we know from the pollen in their fossilized feces) or to celebrate a boy's passage into manhood, or a girl's to menarche; or perhaps they decorated the shelter itself to keep away evil forces – or simply for reasons of esthetic pleasure! The possibilities are many, and one thing we should avoid is to underestimate the subtlety of the minds at work here.

One physical trend through the time of *Homo erectus* was a reduction in the size of the teeth (and consequently of the jaw, with the muscles required to operate it). This has been interpreted to mean that with

The remains of more than one hundred individuals of Neanderthal Man have been found over much of Europe and parts of the Middle East. Because so many fossils have been found in Europe, and well publicized, he has been fixed in the public mind as the archetypal 'brutish' Stone Age Man. The reconstruction shows him to be heavier and stockier than modern man, but he was as effectively bipedal as we are although earlier workers thought otherwise. The taxonomic place of Neanderthal Man is somewhat uncertain: most workers consider him to be a subspecies of modern man.

the advent of cooking, food became less tough and demanded less effort in chewing. All of which, though it sounds reasonable, raises two further questions about early humans: first, why they were interested in fire; and second, why they began to cook their food.

The first is probably easier to answer. There must have been some immediate and very practical benefits to controlling flames. Early humans must have seen fires sweeping through bush and forest following a chance ignition. A smoldering ember retrieved from such a blaze was probably the source of a hearth fire in many an early hominid camp. Warmth and light are two substantial reasons why our ancestors might have attempted to control fire for their own use. Even in tropical Africa the nights can be chilly, especially high up in the hills. By effectively extending the day, a campfire provides a unique social focus, enhancing the interactions of an already very social animal. As *Homo erectus* moved northwards into colder climates, the need for warmth would of course have become more pressing.

But there is unquestionably something sensual about fire, something magical even. Few people can have failed to have experienced the lure of a blazing, crackling bonfire. The sight, the sound, the smell, and the waves of warmth combine in a kaleidoscope of sensations to be almost totally transfixing. The fascination excited by fire is obviously deep within us. Did the people in the Terra Amata camp feel this as they sat around their hearth at the end of a spring day on the shores of the Mediterranean? We can be fairly sure that fire would have been a central element in early ritual.

If fire has a mystical element, so too does cooking. Whether the first piece of cooked meat was produced by accident, when someone's supper fell into the campfire, or as an experiment, just to see what would happen, the one advantage would have been an 'improved' taste; the same would be true for a meal of tough roots or nuts. But if your jaw is already equipped for powerful chewing, there is no hardship in grinding your way through nuts and gristle. And would the changed taste have seemed that much more desirable? At any rate, the benefits would seem marginal. Is it possible, however, that through cooking their food, rather than eating it raw as other animals did, early humans may have been emphasizing their specialness in the world, in culinary declaration of the spirit of humanity?

This interesting skull was found by some villagers, who broke into a cave near their village of Petralona in the Khalkidhiki peninsular, northeast Greece. It has not yet been fully studied, but present findings suggest that its characters place it between Homo erectus *and* Homo sapiens.

Exactly when cooking became a routine habit is impossible to pin down, and the origins of the controlled campfire still lie in a shadowy region of prehistory. Certainly, by half a million years ago the hearth was part of our ancestors' camp life. And there are definite signs of a hearth in the Escale cave in the valley of the River Durance near Marseilles, an occupation site that may be close to a million years old. There are signs, too, of heat-charred stones and earth on a campsite east of Lake Turkana that goes back some two and a half million years. Whether these are the remains of a deliberately made fire while the camp was occupied, or are merely the result of an accidental fire which burned on the site after it had been abandoned, we cannot be sure.

Perhaps the earliest indication of some stirring in the human mind that expressed itself in ritual comes from a limestone-filled cave at Choukoutien, about thirty miles from Peking. At least half a million years ago, the bodies of fifteen individuals were buried there. When they were discovered, in a series of excavations between 1926 and 1941, the skeletons were not intact, and indeed many of the limb bones were shattered. So too were some of the skulls. But in a number of them the opening through which the spinal cord runs had clearly been enlarged – a difficult task, calling for particular care. The scene has been interpreted by some as the remains of a ghastly cannibalistic feast, and thus as sure evidence of man's aggressiveness. That someone went to a great deal of trouble to extract and thus presumably to eat the brains of their deceased fellow beings seems undisputable. But to associate such an act of cannibalism inevitably with aggression is grossly to oversimplify – as we shall argue presently in more detail.

Yes, Peking Man probably did eat the brains of his fellows. Our point here is that the feast would have been a ritualistic one rather than mere gorging on human flesh. Why take such trouble to widen the opening at the base of the skull when it would have been so much easier to smash the cranium and scoop out the soft contents? Whether the aim of the feast was to gain power over enemies by devouring the brains of the vanquished, or to maintain a bond of continuity with a deceased relative, is beside the point. The Choukoutien bones have been only tentatively dated at half a million years; there is a chance that they are considerably older. And in any event we should consider the antiquity of the cultural fabric that links together the meager remains of *Homo erectus* that have turned up throughout Europe, Asia, and Africa.

We noted in Chapter 2 (p.32) the remarkable discovery by Eugene Dubois of what he called *Pithecanthropus erectus*, as long ago as 1891. The young Dutchman had decided that the origins of man were to be found in the East Indies; and it was in Java, among exposed strata at a bend in the River Solo, that he unearthed the first specimen of what was subsequently renamed *Homo erectus*. By comparison, the remains of *Homo erectus* at Choukoutien are much stockier than their counterparts in distant Java. But

wherever they were found, for half a million years – down to approximately 500,000 years ago – these human creatures carried on much the same kind of existence, that of an accomplished hunter-gatherer. These were people in tune with their environment, exploiting every part of the plant and animal kingdoms the season of the year dictated. A number of their encampments show seasonal reoccupations, a pattern that becomes increasingly entrenched as we move closer to modern man. The seasonality of migrating herds, of succulent young plants in springtime, of ripening summer fruits and late-maturing nuts – these are what from week to week governed the lives of our ancestors. We must not forget, of course, the ever-present garnish of opportunities that are the spice of human life.

An American researcher, Richard MacNeish, and his colleagues have reconstructed the seasonal activities of the El Riego hunter-gatherers who lived around ten thousand years ago in the Tehuacan valley of Mexico. During the spring they split into family units who foraged over a wide area for last year's seeds, pods, and the year's new shoots. When the months of summer abundance came round the families aggregated in the valley bottom to collect early fruits and seeds. Then, as autumn advanced, the bands moved up the slopes, where they lived mainly on fruit but also did a little hunting and trapping. Finally, during the winter months they moved back to the valley bottom, where hunting and trapping became their staple means of subsistence. Much the same seasonal pattern must have been followed in every corner of the globe where early humans thrived.

Left: the digs at Choukoutien (Dragon Bone Hill) – about thirty miles southwest of Peking – have provided many fossils of Peking Man (Homo erectus). *Above: a skull is shown being freed from the stratum in which it was found. Although the human fossils are important, equally important is the evidence, from the charred bones of several animals, that Peking Man had learned to make and use fire for cooking and hardening the bones for use as tools.*

Overleaf is an artist's impression of such a scene.

The Dutch archeologist, Eugene Dubois, discovered the first remains of Homo erectus *at a site on the River Solo, near Trinil in Java (now Indonesia). The femur and skull cap of Java Man, as it was called, are shown here. Since the 1890s, when Dubois was working there, many other fossils have been found.*

There is evidence that our ancestors also occasionally conducted spectacular hunting expeditions. For instance, at Torralba in Spain, Clark Howell, an anthropologist from Berkeley, California, found the remains of a large number of elephants, horses, oxen, rhinos and red deer. Howell suggests that three hundred thousand years ago human hunters may have banded together to drive these animals into a swamp where they could easily be killed. There are signs that the hunters lit fires in the bush so as to stampede their prey. The same thing appears to have taken place at Ambrona, just a couple of miles up the same valley. Here elephant bones are most common, with the remains of almost fifty individuals concentrated in a relatively small area.

Probably several bands of hominids would have combined for these hunting drives, if for no other reason than the embarrassing superabundance of meat that even one prey animal would yield. Almost certainly such hunts were seasonal and opportunistic, and do not represent the normal week-to-week pattern of subsistence.

Around the time that the Terra Amata and Ambrona hunter-gatherers were embarking on their occasional slaughters, the Old World could be divided in two according to the basic tool technology that predominated. At Ambrona, for instance, the meat of the slaughtered elephants would have been hacked away from the limbs with Acheulian tools, whereas in parts of India, China, Burma, and Southeast Asia a pebble tool technology rather like that of the Developed Oldowan industry, was in use. This division can presumably be traced to the pattern of migration by early *Homo erectus* bands from Africa. It seems logical to suppose that the first *Homo erectus* people to leave Africa did so before the emergence of the Acheulian culture, some one and a half million years ago, and that they took up residence in the east. Later waves of emigrants would have taken with them the apparently more sophisticated Acheulian tool kit, and have headed west. On the other hand, since it is very probable that contemporary populations of *Homo erectus* in Africa were divided into 'tribes' according to whether they produced the Acheulian or Oldowan technologies, this explanation may be somewhat simplistic.

Whatever accounts for the division of the two cultures in the Old World, that division began to erode at least half a million years ago, and a brief period of dominant Acheulian technology was followed by a proliferation of stone tools which reflected the growing expansion of culture. This initial change may have been because of a steady migration of peoples from west to east. Or, as possibly happened at Olduvai, it may be that the culture spread from one band to the next through social and other contacts, while the populations themselves stayed more or less where they were. On balance it is a reasonable guess that the notion of a mass migration by people bringing superior tools and sweeping all before them is somewhat exaggerated. By definition, cultural entities are highly mobile in themselves, requiring only that neighboring people should be in contact with each other, not that they should migrate. Witness the waves of fashion, such as the craze for denim that swept through the western world in the mid-1970s.

The evolutionary momentum that propelled *Homo erectus* towards *Homo sapiens* must have left behind some pockets of *erectus*, and have caused some *sapiens* populations to move down biological blind alleys to their own extinction. Joining Neanderthal Man on his ill-fated journey were Solo Man and Rhodesia Man, both of whom likewise paid the penalty of over-specialization. For a time the genetic pool of basic *Homo sapiens* must have contained quite a mixture. For instance, around one hundred thousand years ago two individuals died on a flood plain that is now sliced through by the Omo river in southern Ethiopia – one of them thoroughly modern, the other having a skull with many *erectus*-like characteristics. A similar situation has been noted in Israel, where two caves on Mount Carmel have yielded two separate sets of fossils. One set, around fifty thousand years old, was typically modern; the second, older by no more than ten thousand years, appeared to have many archaic characteristics, including the heavy brow ridges and the general form of a classic Neanderthal.

Eventually there emerged the variant of *Homo sapiens* to which every man, woman and child living today belongs: that is, *Homo sapiens sapiens*. To ask where the final transmission occurred, or precisely when, is simply not a sensible question. There is no single center where modern man was born. It is worth reflecting, however, that during the period of the evolutionary emergence of modern humans there were probably at least five times as many hominids living in Africa as in the rest of the world. The relatively large number of archeological sites in Europe from this period is the result of a combination of better preservation of human remains, ease of discovery, and manpower involved in the search. Europe is certainly the focus of recent human archeology, but it cannot necessarily claim to have been the focus of our recent evolution.

In any event, it is probably somewhat misleading to refer to a specific *focus* of human evolution. More likely, in the volatile mosaic of the basic *sapiens* gene pool the transition to *Homo sapiens sapiens* occurred in many different places, the populations at these foci then spreading out to dominate their local geographical areas and eventually meeting similarly advanced populations from other areas. For a graphic analogy of the gene pool, one can think of taking a handful of pebbles and flinging them into a pool of water. Each pebble generates outward-spreading ripples that sooner or later meet the oncoming ripples set in motion by other pebbles. The pool represents the Old World with its basic *sapiens* population; the place where each pebble lands is a point of transition to *Homo sapiens sapiens*; and the outward-spreading ripples are the migrations of truly modern humans. This image is in contrast to the misleading idea of one pebble landing in the pond, and of the ripples that eventually cover the whole surface arising from that single source.

It should be emphasized that the physical variations we see today among people from different parts of the world are variations *within* the subspecies *Homo sapiens sapiens*. Every human being on the earth is a member of that same subspecies. The variations we see come about through geographical separation and through adaption to particular local conditions. For instance, Eskimos are stocky, a physical type best suited to conserving heat. Compare this with the tall, elegant Maasai, whose bodies, like those of many tribes in tropical countries, are well adapted for losing heat. Skin pigmentation is another example, the degree of pigmentation increasing as one moves closer to the equator. As the role of the pigment, melanin, is to protect the skin against ultraviolet radiation, this transition is biologically sound.

The need to protect the skin through pigmentation arose, of course, as the early hominids lost their thick covering of hair. We still have as many hairs as our ape cousins, but they are fine and short and therefore leave the skin virtually naked. There must have been some advantage associated with our ancient 'unclothing', and one strong possibility is that it allowed us to elaborate a very efficient cooling system, in the form of more than five million tiny pores – the openings of sweat glands – scattered all over our body. By evaporating moisture through these pores we can lose heat at a rate not matched by any other animal – a great advantage for bursts of intense activity under a blazing sun. This benefit carries with it the slight drawback of an increased dependence on water: moisture lost through sweating has to be replaced by frequent drinking. Hence humans became more than usually water-dependent animals.

The loss of thick hair probably happened at an early stage in the development of *Homo erectus*, while our ancestors were still in Africa (according to an intuitive guess, since there can be no direct evidence). In that event, increased pigmentation must also have arisen then. But as people moved into colder climates the pigmentation would have become a disadvantage because it prevents what little sunlight there is from catalyzing an essential chemical reaction in the skin that produces vitamin D. So, as some of our ancestors migrated into the cooler northern hemisphere, their skins would have become lighter and lighter through genetic selection. The *Homo erectus* populations who

Oceanics – a Wapenamanda from the highlands of New Guinea

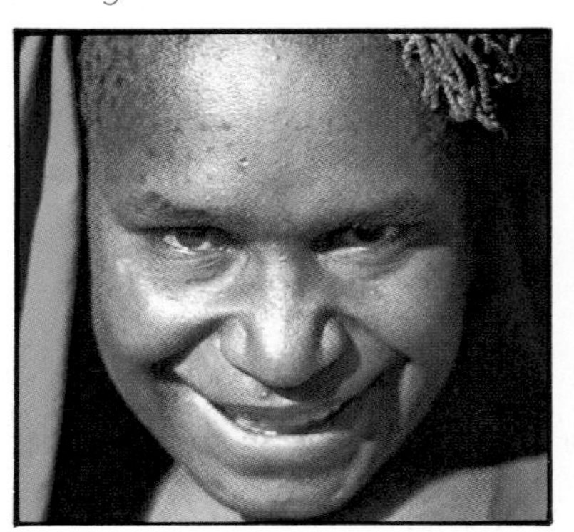

Asiatics – a Chinese from a commune near Peking

Australians – a Lirutji Aboriginal, Northern Territory of Australia

Africans – a Samburu from the northern area of Kenya

The physical variations among people from different parts of the world come about through geographical separation and adaptation to local conditions. The geographical races into which Homo sapiens sapiens *is divided are a convenience, for there are no clear-cut physical distinctions. Cultural differences are rather more distinct.*

Skin pigmentation is an example of an adaptation to the environment. The diagram shows the granules of melanin in the germative layer, and it is the density of these that are characteristic of each race.

Negroid

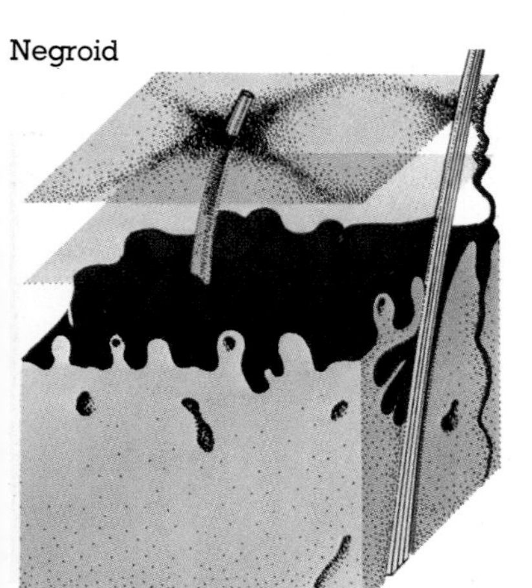

Caucasoid

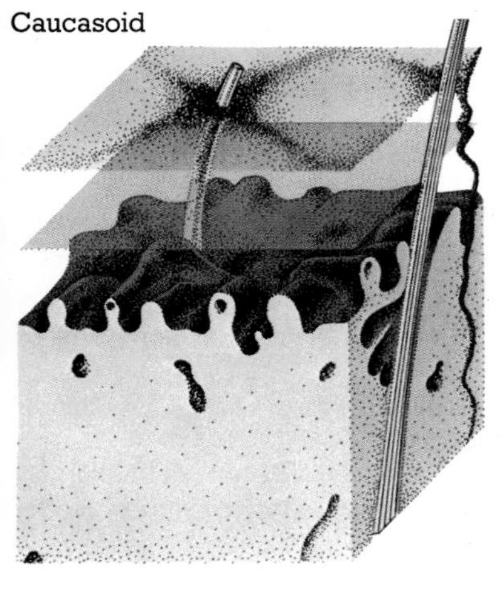

Mongoloid

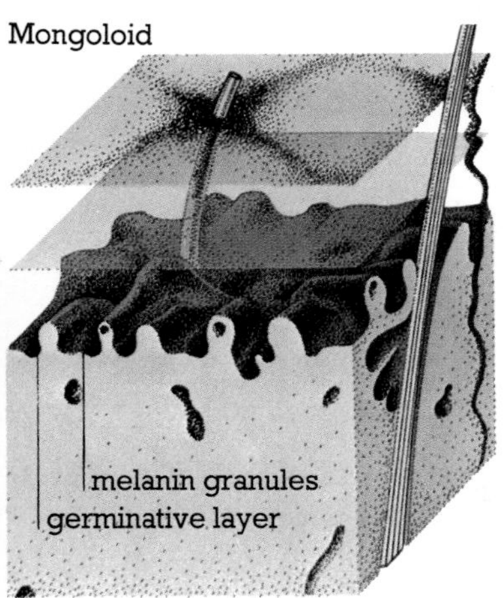

Amerindians – a Quechua Indian from Peru

Caucasians – a nomad chief from Afghanistan

The 'laurel leaf' flint 'tool' tells us, as do the cave paintings, that Cro-Magnon Man was possessed of great artistic skill. Because it is so delicate it is probable that it served no useful purpose but is an expression of fine craftsmanship and sense of form.

Because of the climatic conditions in Africa people did not need caves for shelter and so of the paintings done on rock faces relatively few survived. However, those of more recent date have done, as at Tassili in southern Africa, where there are many examples of Neolithic cave art.

remained in Africa, and some of whose descendants eventually also made the transition to *Homo sapiens sapiens*, would have remained dark-skinned.

It should not, therefore, be surprising that as light-skinned men were executing elaborate and beautiful cave paintings in Europe, equally sophisticated art was being produced by their darker-skinned brothers in Africa. Because the people living in Europe some twenty thousand years ago were battling with ice-age conditions they generally retreated to caves, where they could escape the worst ravages of the cold. Deep in these caves they painted pictures, usually of animals and frequently depicting a hunt. We can imagine that these people had a deep respect for their prey animals: as expert hunter-gatherers they would have had as intimate a knowledge of the living world around them as any biologist today; they would not have been unaware of the complexity and subtlety of the animals' lives. The paintings may well have been a device for ensuring successful hunts, but they may also have been an expression of the spiritual conflict involved in the destruction of life. It is surely not insignificant that plants do not figure greatly in pre-

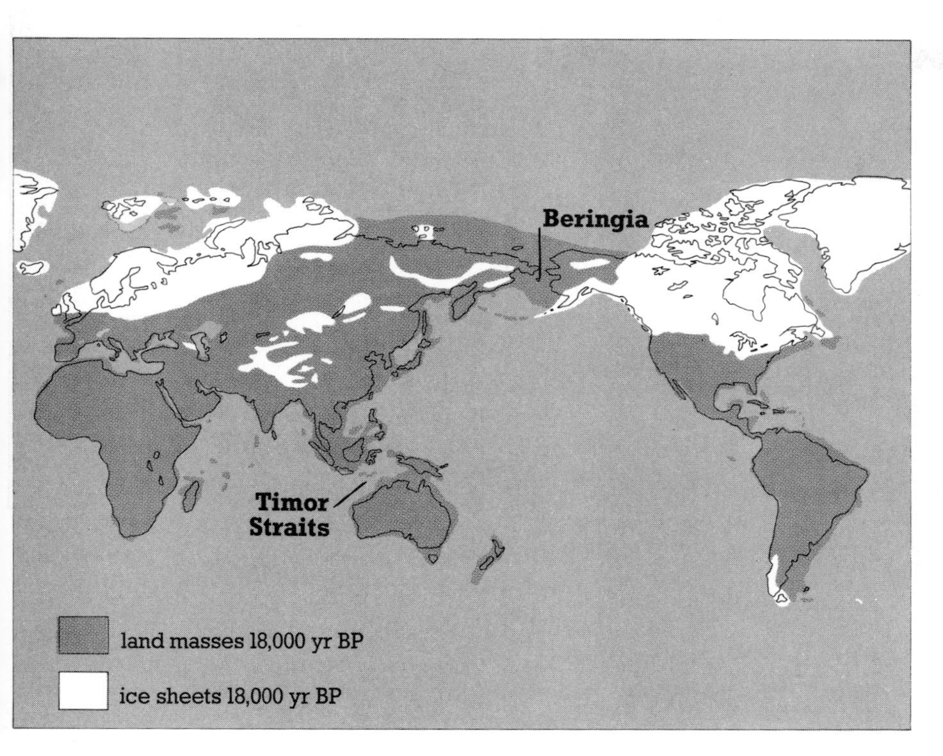

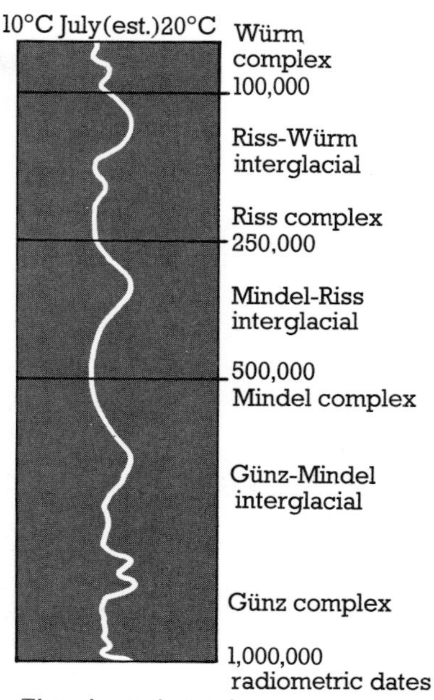

The chart shows how the temperatures have fluctuated during the last million years.

The advance and retreat of the glaciers during the last Ice Age played an important part in the dispersion of Homo sapiens. *The map shows the maximum spread of the ice sheets – about eighteen thousand years before the present. At this time the sea level would have been considerably lowered and thus land bridges would have been created between the North American continent and Asia – and the straits between southeast Asia and Australia would have been narrowed.*

historic art. Although plants are alive in the biological sense, they perhaps did not emanate a feeling of 'life' in a way to which early humans might respond.

Whatever the motives for prehistoric painting, and there must have been many, the sheer delight of artistic skill and esthetic enjoyment must have become important too. Perhaps the surest sign of an esthetic sense from this period comes in the form of a stone 'tool' known as a 'laurel leaf'. The best such work of art, for that is what it really is, was discovered in south-eastern France in 1873: shaped like a slender leaf, it is fourteen inches long, four inches wide at the center, and a mere quarter inch thick! Used as an implement, it would certainly have shattered. No, these blades are expressions of fine craftsmanship and sense of form. It is conceivable, of course, that 'laurel leaves' were used as a kind of currency, perhaps forming part of a bride price, just as are finely worked shells in some 'primitive' communities today.

The timing of the last ice age that drove Europeans to live in caves means that many of their artistic efforts have survived the passage of time, and many magnificent examples have been discovered in France (more than sixty known caves) and Spain (so far around thirty). Because of the relatively gentle climate in Africa, people there did not have to take refuge deep in hillsides. This means, unfortunately, that most of the prehistoric paintings there were on relatively exposed rock faces that have gradually flaked away through time. One striking exception, however, remains: in the Cheke district of Tanzania there are a number of rock faces which formed parts of prehistoric rock shelters, painted with scenes of animals and humans in much the same styles and compositions as in Europe. Giraffes and elands pre-

dominate in the paintings, and they are often represented in fine detail. By a curious contrast, drawings of people had simply a circle or a blob for a face, never any detail. There must have been some kind of taboo against representing a person's face, perhaps from the belief that such an act would rob those represented of their spirit.

By the time the people of Africa and Europe were at work on their more advanced paintings, mankind had already conquered two more continents, Australia and the Americas. In both instances there is the problem of how they traveled from the Europe/Asia/Africa land mass to what today are islands. How did they cross the seas? Between about one hundred thousand and ten thousand years ago glaciation increased and massive ice sheets soaked up the oceans like great sponges, lowering the sea level by as much as 400 feet. This was more than enough to expose the Beringia land bridge between North America and the northern tip of Asia, leaving a dry, if rather chilly, route for the journey into the New World.

Australia poses more of a problem, however, as the Timor Straits separating Australia from South-east Asia are just too deep to have been emptied during the glacial period in which men apparently crossed them. The inescapable conclusion must be that the immigrants to Australia sailed there. Although the effect of the glaciation was to narrow the Straits considerably, there would still have been about sixty miles of open sea stretching out in front of those early adventurers when they made their journey, twenty thousand years ago. What could have motivated such a trip is almost beyond imagination. The people who attempted it would have had no firm knowledge that anything existed beyond the visible horizon. They may, of course, have seen flocks of birds flying out to sea, and wondered where they had gone. What would their craft have been like? Dugout canoes in all probability – hardly ideal for a seagoing journey.

If the voyage was indeed intentional – for there is the possibility that an inshore canoe might have been blown out to sea, to begin the adventure unwittingly – then it is a tribute to those people's spirit and courage.

Cro-Magnon Man was, in lhe anatomical sense, truly modern – Homo sapiens. *The fossilized remains are identical with those of people living today.*

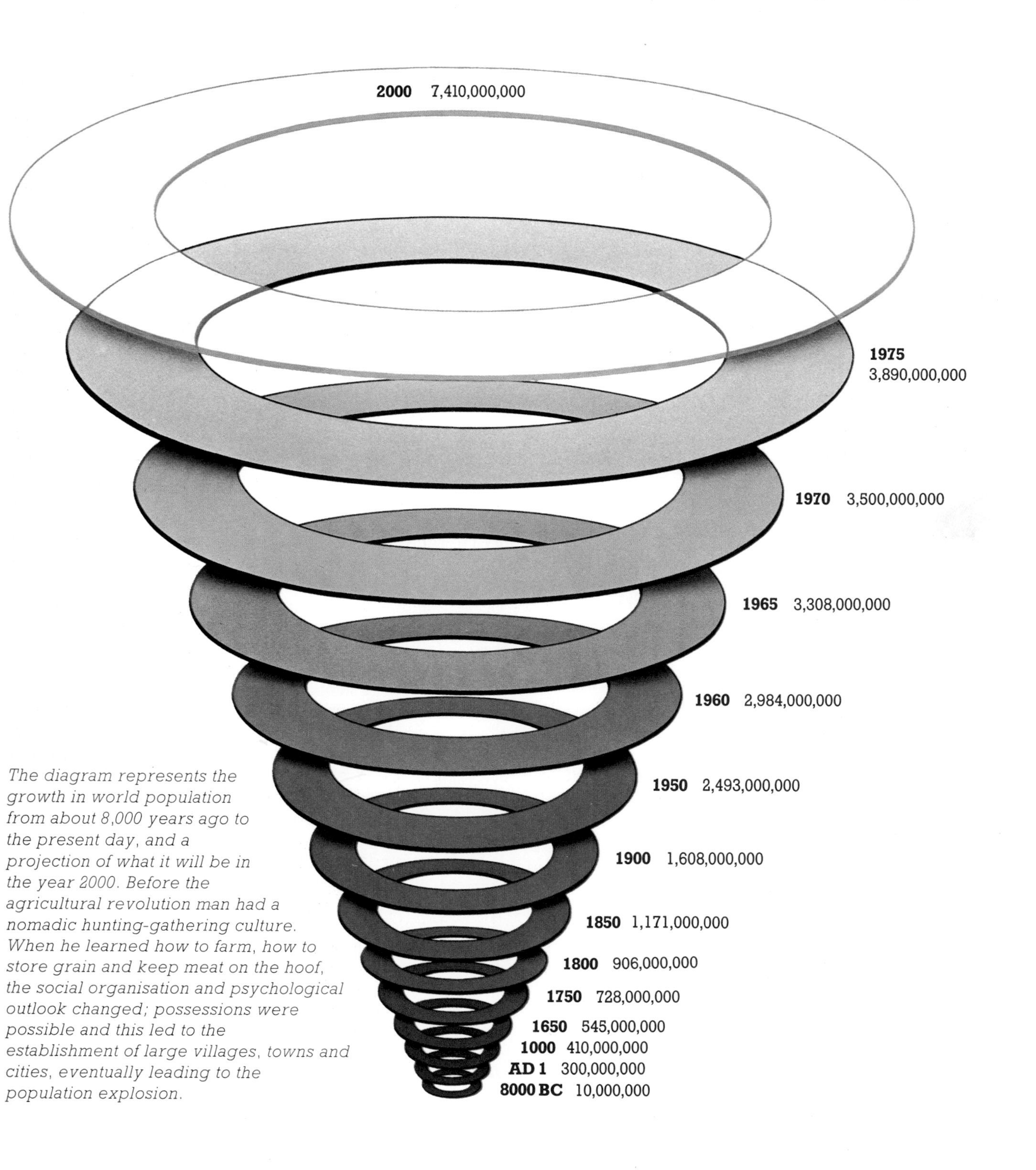

The diagram represents the growth in world population from about 8,000 years ago to the present day, and a projection of what it will be in the year 2000. Before the agricultural revolution man had a nomadic hunting-gathering culture. When he learned how to farm, how to store grain and keep meat on the hoof, the social organisation and psychological outlook changed; possessions were possible and this led to the establishment of large villages, towns and cities, eventually leading to the population explosion.

However it came about, it must have happened more than once: for although the early prehistoric remains from this period are not superabundant, there are enough to indicate that more than a single canoe load of individuals were thriving in Australia eighteen thousand years ago.

Although the migration into America presents less of a puzzle in terms of method, there is, apparently, an issue of timing: with too little glaciation the land bridge would not be exposed, and with too much the route into North America could well have been blocked by a huge wall of ice. The amount of glaciation, therefore, had to be just right in order to expose the dry route and still leave an ice-free corridor into the New World. Geologists are still in dispute over when such a propitious state would have obtained; there are suggestions that around twelve thousand, twenty-five thousand, and seventy thousand years ago would have been suitable times. Meanwhile, archeological evidence scattered over both North and South America points to an earlier rather than a later date. Although most sites are younger than twenty thousand years, a skull from what is called Los Angeles Man (because that was where it was found) has been dated at about twenty-six thousand years ago. As there is no good reason to suppose that once the first people arrived in the New World they were hellbent on reaching Los Angeles, one can infer that the predecessors of Los Angeles Man stepped onto the shores a *very* long time before he lived and died.

The people who had made the trek across the Beringia land bridges would have been hunter-gatherers who must have already learned to cope with arctic conditions similar to those which the North American Eskimos now face. It is tantalizing to speculate on whether the basic *Homo sapiens* people of northern Asia and Europe during the previous great glaciation – some two hundred thousand years ago – might not also have been able to make the journey into the Americas. There does not appear to be any powerful reason to reject the idea, in which event the American Indians would have had a much longer ancestry than previously supposed. In any event, some recent excavations in Canada and Mexico are uncovering signs of human occupation dating back as much as fifty thousand years. It may be that sites four times that age are just waiting to be discovered!

By about ten thousand years ago, with virtually every part of the globe populated, however sparsely, humanity was in place for the advent of agriculture. At the time hunting and gathering was the universal means of subsistence, each band of humans exploiting the seasonal offerings of the animal and plant kingdoms in its own locality. By now the bow and arrow had been invented, as had the spear and spear-thrower; both of these were important technological advances for hunting. The technology of plant- and food-gathering, however, remained simple: merely a container in which to carry fruit, nuts, and succulent roots back to the camp. Life was essentially nomadic, unhurried, leisurely.

In general, hunting and gathering bands were relatively small, consisting of perhaps five or six family units. They would be part of a large and widespread tribe, sharing the language and culture of their neighbors but subsisting as a small, mobile band. Some hunter-gatherers, however, did not have to move camp every few weeks in search of new food sources. Some even built small villages, containing a hundred or more people. The reason for this unusual stability would have been a particularly rich food source. One such village is in Lepenski Vir, Yugoslavia. There, on the eve of the agricultural revolution, a band of hunter-gatherers built a village perched above the rushing Danube. Although they gathered plant food from the surrounding countryside, their main subsistence was on fish from the river. In their village they carved faces on boulders, giving them distinctly fishy expressions.

Another hunting and gathering village dating to the same era is Ain Mallaha in the Upper Jordan Valley. Here, the staple food source that supported the village belonged to the plant kingdom: pistachio, oak, wild wheat, and wild barley. All the villagers had to do was to go out with their stone-bladed sickles and collect or pick what was ripe. They hunted too, gazelles being common in the area. Such was the success of this arrangement that the village housed at least two hundred people, and perhaps as many as three hundred. It must have been this kind of fortuitous combination of an abundant supply of plant food and a sufficient supplement of wild meat that originally gave rise to controlled cultivation and farming.

An obvious first step from the straightforward gathering of abundant plant foods towards actually cultivating them is simply to help them grow a little better – by irrigation, for instance. Until not very long ago the Paiute Indians of the Owens Valley in the south-western United States did just this. They dug irrigation canals which they fed from dammed streams to enhance the growth of their plant foods, none of which they planted themselves. Taking care of growing plants is certainly a step towards agriculture, but the distinctive element is the actual sowing of the seeds, and even more the sowing of specially selected seeds. The backbone of successful crop-growing has been genetic selection for high yield and for resistance to diseases. For instance, the cob of the maize that is grown today is at least ten times larger than that of the original species from which it is derived. Not only that, the cob in modern maize is tightly enclosed in overlapping husks, thus preventing the possible loss of any kernels during collection.

Maize was one of the first crops to be cultivated, and the initial steps in the selective breeding towards today's super-cob were probably fortuitous. The simple act of gathering the tiny wild cobs would tend to select those in which the kernels fall out least readily: if the cobs are taken back to the village to be dried, the kernels that survive the journey will be those which stick most tenaciously in the cob. Once people had taken the conscious step of sowing seeds it was then just a matter of experience and insight to improve the crops by using seeds from the healthiest plants of the previous season.

Whether, initially, deliberate sowing followed a conscious experiment, or was the result of keen observation of accidentally spilled seeds that had been meant for food, is a matter of history that we shall never know. In any event, just as there was no single focus for the beginning of *Homo sapiens sapiens*, so there was no one spot where agricultural crops were first established. The event must have occurred independently in many places and in many different ways, sometimes perhaps accidentally and sometimes by design. And the circumstances for beginning regular root-cropping as opposed to seed-cropping must also have differed. The previous focus of any hunting and gathering community must, of course, have influenced the style of agriculture it adopted.

In this respect there are two extremes of hunting and gathering subsistence, such as would give rise to two extremes of agriculture. For people like those in Ain Mallaha, dependence on wild seeds would have led them to seed-crop farming; others, who concentrated on exploiting migrating herds, perhaps following them for long distances, might have taken to husbanding the animals. An intermediate stage for this second group might have been a pastoral existence, herding their animals across the countryside rather than keeping them in defined areas. In their seasonal wandering with their animals, pastoralists can, of course, plant a few crops to supplement their economy, just as the Dassanetch people do today.

Although most of the emergent farming communities would not have depended on one single crop, their different specialities would have encouraged them to trade with one another, exchanging grain for meat, for instance. Trade was not a new idea in the world of ten thousand years ago: obsidian and other materials for making stone tools had already been the subject of exchange many thousands of years before. Nevertheless, with the growth of agriculture there would have come a new and increasingly urgent need for the exchange of goods.

Why agriculture began when it did is difficult to say. Perhaps, though, it was a combination of the onset of balmier conditions with the passing of the ice age, and the social readiness of the few sedentary hunter-gatherer communities. In any event it began in many areas, though perhaps the most favored locality was the fertile crescent around the southeast edge of the Mediterranean, where the archeological evidence tells us that the new habit spread quickly and was soon supplying all the food needs of the communities. This contrasts with Meso-America, another focus of agriculture around ten thousand years ago, where plant cultivation went hand in hand with hunting for several thousand years. It was not until a little over three thousand years ago that maize cultivation, combined with that of squash and beans, could support stable communities. Other early centers for agriculture were Peru and Thailand, and no doubt many yet-to-be-discovered archeological sites will show that the list of such centers is actually very long.

Once the agricultural revolution had begun, this new form of subsistence spread with extraordinary rapidity. Because food supplies could be concentrated into small areas, large villages, towns, and cities became possible. The world's population shot up from around ten million at the beginning of the agricultural revolution to a figure that is now over four thousand million. Only a tiny fraction of this staggering number now live by hunting and gathering, an economy that for over three million years not only provided our ancestors with subsistence but also shaped our social, psychological, and physical evolution.

7 The First Mixed Economy

For a huge slice of our evolutionary career we were hunters and gatherers: it was a mere 10 000 years ago that the first people began to experiment with the possibilities of organized agriculture. Although killing animals provided a new source of food for our ancestors, this utilitarian aspect of hunting pales into insignificance compared with the enormous impact that the mixed economy of hunting and gathering exerted on the march of human evolution: it shattered the basic primate social pattern, and replaced it with a uniquely human society based on division of labor between individuals; it enhanced pressures selecting for more brain power; and it demanded a degree of social cooperation not displayed by any of our primate cousins. The key to the first mixed economy was sharing.

Because the archeological record is so tantalizingly sparse, we cannot yet say exactly when meat-eating became important to human life. But it is reasonable to suppose that meat-eating, either scavenged or obtained by hunting, was a factor, and possibly the major one, in the gradual emergence of the *Homo* line from the basic hominid stock, leaving a predominantly vegetarian niche to be occupied by the australopithecines. It may even be, of course, that at least limited meat-eating had something to do with the new ecological niche in which the hominid stock became established in the first place, some twelve million years ago. But the rise of a hunting and gathering existence is more likely to have occurred roughly five million years ago, along with the emergence of our direct *Homo* ancestors. And it may be that large-scale big-game hunting, which our ancestors occasionally engaged in, developed no more than two million years ago.

Whatever may have been the precise timetable for the adoption of a hunting and gathering way of life, it is certain that, more than any other behavior, it is to this social pattern that we must look for clues to any *basic* human characteristics. Mankind hunted in a big way for at least a million years, and subsequently farmed for ten thousand. We can be sure that it was the intellectual equipment and the social cohesion fostered by the hunting and gathering way of life, reinforced by the social pattern of sharing, that made organized agriculture possible at all. In the same way, the industrial revolution of the nineteenth century and the technological wizardry of the twentieth alike owe their existence to our hunting and gathering heritage. During the ten thousand years since the hunting life was largely abandoned, there can have been no significant biological changes in the human animal – the time would have been too short. Much of what makes us human is in us because we developed the unique habit of collecting and sharing plant and animal foods.

Of the many myths that have passed for theories concerning human evolution, those concerned with hunting have been among the most highly colored, and the most dangerous. The fact to be remembered is that there is a distinct and unbridgeable gulf between hunting and aggression. To speak of a 'primeval lust for flesh', or the 'blood-bespattered archives of human evolution', or in other such emotive and equally inaccurate phrases is, biologically, total nonsense. And the nonsense becomes pernicious when it is used to justify the apparent proclivity of modern humans for exterminating one another through the ever more sophisticated technology of warfare. Our own concern in this chapter, however, is with what hunting and gathering imply for the social structure of a meat-eating animal.

Our close primate cousins are almost without exception largely, though not exclusively, vegetarians. Even those that stray into meat-eating – the chimpanzees and baboons – do so only very infrequently. Primates are social animals, and this is one element of their success. But a plant-eating existence tends to make the individual members of a troop very self-centered and uncooperative. In spite of the intense social interactions that take place, especially in our closest relatives the chimpanzees, and even though there is group awareness of a sort in the search for food, to be a vegetarian is to be essentially solitary. Each individual tears the leaves from a branch, or plucks fruits from a tree, and promptly eats them. There is no such thing as communal eating or the sharing of food among our close relatives.

There are exceptions however, and these would appear to be significant. For instance, when chimpanzees have caught a young baboon or monkey they will share their spoils. But even so, the sharing is not *active*: the successful hunters do not return to a base

Previous pages: The G/wi are contemporary hunter/gatherers living in the Central Kalahari. Anthropological studies of their way of life, and that of the !Kung, have revealed much that might throw a light on our early ancestors' lives.

camp and divide the meat among other members of the troop. A non-hunter hoping to obtain some share of a recent kill must beg pretty consistently before a morsel is handed over. Baboons, the only other meat-eating higher primates, also share, but to an extent much more limited than in the chimpanzee.

When our ancestors took up organized hunting and gathering as a career, therefore, one of the major changes in their way of life, and one which was to open up a vast behavioral gulf between humans and our closest relatives, was the adoption of sharing. This new and unusual form of primate behavior was one of a whole group of traits acquired through hunting and gathering that helped to push our human ancestors towards an increasingly adaptable way of life. And it was adaptability that enabled the human species to thrive in practically every corner of the globe. Basically, we can suppose the package of social traits derived from being a hunting and gathering animal to be something like this: a base camp, where infants could be cared for and to which both meat (which might have been actively hunted or opportunistically scavenged) and plant foods could be brought became an important social focus, a division of labor in which the males hunted and the females were responsible for child care and for gathering plant foods; and the development of a tremendous need for cooperation and restraint, every individual being more dependent on the activities and trust of others in the group than ever before in the world of primates.

Hunting and gathering reshaped the lives of our ancestors in other ways as well. For instance, hunting hominids ranged over a vastly bigger territory than any other primate, covering more ground in one minor hunting trip than most primates do in a lifetime. Differences in physique between hunting males and gathering females were accentuated. Sexual activities became more controllable, and possibly more meaningful too. But overall, the three most important effects of hunting on the lives of our ancestors were the establishment of a base camp, the division of labor, and the ensuing cooperation. These in turn allowed a closely knit social group to form, in which prolonged education of infants was possible, an education that was necessary to equip individuals with skills they needed to participate in the complex social environment and to contribute to the group's economy, either by hunting or gathering.

Even in the behavior of *Ramapithecus* there had been signs that the period of childhood dependence was becoming prolonged. By the time that *Homo erectus* made its appearance, around one and a half million years ago, the young appear to have been almost wholly dependent for at least the first six years of their life, and possibly eight, much as children are today. This compares with about one year for monkeys, and three or four among the apes. At least part of the reason for a lengthy childhood in humans is the crown of evolutionary achievement we carry in our heads – the human brain. An uneasy compromise has been evolved, balancing the protection necessary for the delicate developing brain, especially in its early stages, against the safe delivery of the baby's head through its mother's pelvis.

A brain that was fully developed at birth would escape the potential hazards of environmental

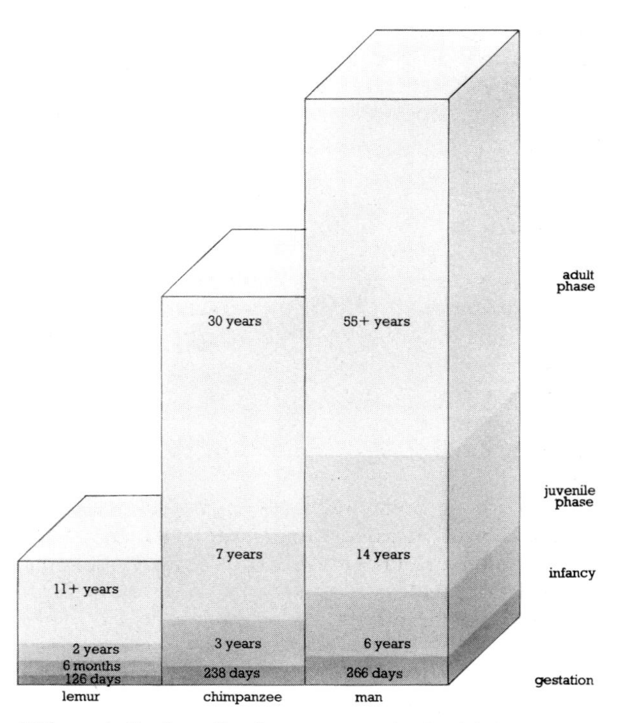

Although the length of pregnancy in the higher apes is much the same as ours, man has a longer infancy phase – that time during which the young is wholly dependent upon its mother. This period becomes longer and longer as one progresses up the primate order and man's is double that of the apes.

Left: The staple diet of most primates is vegetable food and the chimpanzee (left) is no exception. Also, like many other primates, it feeds on insects but, significantly, it does eat meat and this has led to cooperation and sharing. The baboon (below) also eats some meat, which is shared, but to a limited extent.

Right: A fire reinforces the social bond: it becomes the focus of a group and allows contact to be maintained during the hours of darkness, as here where a !Kung trance is in progress.

damage before its development had been completed. But such a brain would inflict impossible demands on the engineering of the birth canal in the mother's pelvis, making upright walking a problem: in place of the characteristic swivel of female hips we know now, a pelvis with a bigger birth canal would impose a marked waddle. A brain delivered into the world in a very small, underdeveloped state would, however, be too vulnerable. The path taken by the evolving human structure was a middle one: an infant with a relatively well-developed brain is delivered through a canal that has caused the pelvis to widen as compared to the male's, but not so much as severely to impede walking. (Women in general, however, will never run as fast as men because of this wider pelvis.)

At birth a human baby's brain is about one third of its

adult size. However, all the building materials – the nerve cells – are in place. What happens in a growing child's head over the first few years of its life is that the crucial connections between the nerve cells are established – a process to which both the amount of food the baby gets and the amount of stimulation in its environment are crucial. During these early stages, too, emotional links with the social group are forged. The importance of these ties is demonstrated in practical terms by the marriage patterns in contemporary hunter-gatherer societies. Marriages are almost always arranged between separate bands, and initially the male leaves his own band and joins that of his bride. But this arrangement continues only until the man has demonstrated his ability to support her, or to produce children, whereupon the couple return to the male's own band. This custom must have something to do with the young hunter's familiarity with his home terrain, and also with the close relationships that will have developed among the men as members of a cooperative hunting team.

These close ties are just one aspect of the stable, cooperating social group that is created by a hunting and gathering way of life. Through our evolutionary career it is clear that any factor tending to bind individuals more closely together contributed to the eventual success of the species. It is therefore intriguing to contemplate the effect of fire in reinforcing the social bond. Not only did it provide warmth, but it also stretched social intercourse into the hours of darkness, a time when the hearth would be the focus of the social group. So, as the flames were keeping potential predators at bay, they were also drawing people together, giving an opportunity for telling stories and creating myths and rituals, as well as for the more mundane task of planning the following day's activities – such as, who would try their luck at hunting, who would form a band to go out and gather several day's worth of plant foods, and who would stay behind in the camp. The importance of spoken language in all this should be obvious.

It is, of course, in the behavior and social patterns of our hunting ancestors that we are most interested: how they organized their hunting and gathering economy, and what this implied for their quality of life. We know that such a society based on a hunting economy was very successful, almost certainly the most successful form of society before agriculture, because our ancestors were able to take their skills to every part of the earth, often under the most hostile conditions. But it would be useful, as well as intriguing, to arrive at a deeper insight into the origins of hunting itself. With due caution, we may do so in two rather different ways: first, by looking at animal 'models' of hunting societies, particularly the African wild dogs, though there may be important differences too because, unlike our ancestors, these animals depend almost exclusively on meat; and second, by observing the customs of contemporary hunter-gatherers. We should remember that these latter are not 'fossilized societies', reproducing exactly the lives of our hunting ancestors, but their social organization does offer some clues about the basic elements important to this sort of society.

Because different species sharing similar ecological niches will probably exhibit similar patterns of behavior, it is possible to look to baboons as a model, however limited, for *Ramapithecus*, our early ancestor. Baboons, rather than chimpanzees, are chosen because their lives are played out against a similar physical environment to that experienced by the first hominids. But since within the primate order, hominids are the only ones to hunt for a living, we are obliged to look outside that primate order for insight into our own hunting behavior. The animals most likely to tell us something about ourselves are the social carnivores – the lion, the wolf, the African wild dog, and the hyena. Like primates, these creatures live in groups, and they also kill for a living. What we want to find out from them is the advantage of hunting in groups and the social consequences of this type of biological economy. We are looking for some very basic principles of the life of social carnivores and although the picture will not be crisp and clear, we should be able to discern some outlines, however sketchy.

Most meat-eaters are, of course, not social: a mouse for instance could hardly be the focus of sharing among one's fellow carnivores. Nor could prey that scurries in and out of the dense vegetation of a forest; a band of predators in pursuit would be likely to end up somewhat bruised, in total disarray. One would thus expect to find truly social carnivores operating in open terrain, in pursuit of large animals. By and large, this turns out to be true. The assumption that hunting in groups is more successful than a solitary foray is confirmed by studies, for instance, of jackals. These creatures often choose young Thomson's gazelles as their prey – an enterprise that is made difficult by the frequently courageous defense exerted by the mother. If a lone jackal tries to kill a young gazelle, its chances of securing its supper are about one in six. But if two jackals share the hunt, they are four times more likely to be successful. Similarly, lions increase their success

Group cooperation enables the wild dogs (right) to bring down prey – such as this zebra – weighing a great deal more than they do individually. Such cooperation also enables social carnivores to select and cut from a herd of wildebeest a weaker member, as the spotted hyenas (above right) have done.

Left: Studying the hunting behavior of the social carnivores can give us an insight into the likely behavior of our ancestors. One such social carnivore is the wild dog.

rate when more than one hunter joins in the pursuit. And among a troop of baboons in central Kenya in which one of the males developed a predilection for meat-eating, the success of the hunt became much greater when other males joined in. Eventually, four or five males actively cooperated in chasing the prey, which frequently was a young gazelle.

Cooperation, therefore, brings its rewards. But those rewards are worth having only if they supply a substantial meal: there is no point in a group or band of predators' coordinating their skill if the end result is a light snack for just one of them. As it happens, group pursuit not only brings more frequent success, but also means that bigger prey can be tackled. For instance, a lone cheetah or leopard thinks twice about taking on potential prey larger than 120 pounds. A pack of wild dogs, which individually are much smaller than the solitary cats, can bring down a zebra weighing more than 500 pounds. In terms of food supply it makes sense to catch as much meat as possible in one foray. The fundamental equation in hunting, however, balances the size of the kill against the effort involved, combined with its probable success and the number of stomachs to be filled. Just as it is uneconomical to invest in a hunt that brings slim returns, so too is it wasteful (in terms both of effort and of meat) if the prey consistently provides an embarrassing surplus.

Group living may also mean the opportunity of muscling in on someone else's kill. For instance, even though a solitary hyena is no match for a lion, a pack of hyenas constitute a threat so impressive that the lion may abandon its hard-won meal. Scavenging, once thought to be the hyena's sole means of survival, is undoubtedly practiced by all the social carnivores. These creatures are, above all, pragmatists, taking the path of least effort for greatest reward. If an opportunity for scavenging presents itself, then it will be exploited. Where the prey is killed rather than scavenged, the predators spend some time choosing animals most likely to be easy to catch: youngsters, the sick and aged, and isolated individuals. Even the weakest and most timid antelope is relatively safe if it remains part of a herd, but once it becomes separated it is doomed to become meat for some nearby carnivore. Those carnivores, if they do not come across isolated animals, have as their first task to separate an individual from the security of its group. There is a lot of evidence to suggest that both wolves and wild dogs have a sharp eye for spotting the more vulnerable members of a herd, an ability they put to good use when they are initially stampeding a herd of potential prey.

A carnivore that is particularly adept at rapidly selecting vulnerable prey will have increased its efficiency enormously: fruitless chases will be avoided. Skill at this stage of the hunt was almost certainly central to success in the hunting exploits of our early ancestors. It is significant that contemporary hunter-gatherers invest most of their skill and knowledge in selecting and creeping near to their prey, rather than in heroic chases or indeed in the sophistication of their weapons.

Hunting in groups therefore has important advantages, but only if the members of the group act in a coordinated way. For instance, wolves pursuing a large prey, such as a moose, usually run behind and alongside the victim, biting its body and legs until the animal is weak enough to be pulled to the ground and disemboweled. During the chase, which may go on for several hours, unrestrained attack would disrupt the more successful strategy of attrition, as well as being somewhat foolhardy. In a pack of wild dogs' pursuit of a fleeing gazelle, not every dog is out to make the first strike. Two or three members of the pack lead the running, and when the prey changes direction the second or third dog may cut the corner of the chase, thus taking up the lead. This kind of chase gives the impression of a relay race, but the relay is a response to the attempts of the fleeing animal to evade the leading dog. Once again, the pursuers attempt to bite the victim's legs and sides until it can be brought down.

When the animal stands firm under attack from wild dogs, a remarkable display of restraint in one of the dogs may be observed. An apparently common technique of securing prey is for one dog to snap hold of the animal's upper lip, thus immobilizing the victim as though it were tethered. Meanwhile, the other dogs disembowel the unfortunate creature, bringing a death which, though appalling to watch, is mercifully swift. The restraint on the part of the tetherer is the more interesting in that the other members of the pack are already beginning to tuck into a meal. Such a dog knows its role, and sticks to it.

Most hunts by social carnivores involve a chase. An exception is the lion, which is too large for prolonged exertion of energy and is therefore obliged to adopt other, more subtle tactics. Careful stalking, leading up

Right: An elaborate system of hand signals is used by hunting people today. The !Kung is giving the sign for a secretary bird. It is likely that similar signals were supplementary to early man's relatively simple language.

Below right: A band of G/wi stalking giraffes in the Central Kalahari.

to a brief chase that sends the prey into the waiting jaws of another member of the hunting pride, is common among lionesses. Restraint and awareness of one's fellow hunters are essential if this technique is to be effective; a premature move could mean the loss of potential prey. Jane Goodall has seen the Gombe chimpanzees organize a trap for a young monkey by stationing themselves at the bases of trees that might offer an escape route for the chosen prey – action that implies lots of restraint, a good deal of cooperation, and more than a little communication.

Means of communication under these circumstances remain something of a mystery, but it can be said that the amount of information individuals can pass to each other – either by vocalizations, gross gestures, or bodily actions as a whole – has been greatly underestimated. When baboons hunt, the members of the troop seem to be guided very much by the apparent intentions of the leader as displayed by his actions. The social carnivores almost certainly depend a great deal on picking up important intentional clues by watching one another – any vocalization being a potential giveaway when prey is being stalked. Indeed, contemporary hunters place a great premium on quietness throughout an excursion, even while the prey has not yet been sighted. Using an elaborate system of hand signals, they can communicate not only the type of animal that has left a track, but also its probable age, state of health, and whether it is in a hurry or merely ambling along.

Having secured their prey, the social carnivores share the spoils, a process that involves more or less fuss. A lion kill, for example, is a very noisy affair, with the dominant males taking first place. Each member of the pride obtains just as much food as it can in the face of snarling confrontations with more dominant members. Even the youngsters must be courageous and persistent to get their fill. By contrast, the wild dogs exhibit something no other social carnivore does: *active* sharing. The hunters bolt down meat from the kill

There is much snarling confrontation among a pride of lions while feeding (above), each member taking as much food from the kill as it can. Even the cubs must be very persistent to obtain enough food. In contrast, the wild dog (left) exhibits active sharing, and a mother will regurgitate in response to her cubs repeatedly thrusting their noses into the angle of her jaw.

on the spot before returning to their den, where they are greeted with great excitement by pups and the caretakers that have been left behind (the mother, who may be accompanied by other adults). Pups and caretakers alike thrust their noses into the angle of the jaw of a hunter, a behavior that usually initiates a prompt regurgitation in response. In a wild dog pack, even sick animals unable to hunt may be sustained for a long time – a very unusual display of altruism in the animal world. This remarkable behavior almost certainly has its roots in the cooperative nature of the wild dog group.

For the most part, social carnivores do not store meat in or around their dens. This is probably a wise precaution, since it tends to protect their young from predatory carnivores that might be attracted initially by the prospect of scavenging. In this respect the early hominids appear to be exceptional. Although archeological remains suggest that very big animals had probably been butchered on the kill site, there are ample indications – especially from two to three million years ago onwards – that bones were taken back to the living base. It may be that the apparent increase in signs of prey on living sites reflects not only a rise in hunting success, but also an increasing respect for the hunting hominids on the part of contemporary carnivores. If, besides hunting, the early hominids made their powers felt by occasionally chasing away other carnivores from their own kills – thus adding to their meat supply by active scavenging – the carnivores would gradually have developed a fear of the hominids. As this happened, it would have become safer for our ancestors to have around the camp pieces of meat that might otherwise have been tempting to such predators. Later, of course, fire would serve that same purpose very well.

It is, perhaps, reasonable to suppose that, because of a shared or at least overlapping food source, and because of continuous confrontations over scavenging, the hominids and the contemporary carnivores would have been less tolerant of each other. Certainly, intolerance appears to be the norm between carnivores of today. To stretch that evidence to the point of inferring open competition between different bands of hominids is, however, unreasonable. Indeed, many of today's carnivores show a high degreee of tolerance to neighboring groups of their own kind. But it is quite clear too that any animal, not just a carnivore, may become aggressive toward any neighbor if the two are under ecological stress – as, for example, when food is scarce. One certainly cannot infer that to be carnivorous necessarily implies aggressiveness.

How would our ancestors of some three million years ago have fitted in with the established balance of predators and prey? We have already noted that our primate heritage makes us daytime animals. It happens that the only other daytime hunters among today's carnivores are the wild dogs, whose forays are taken at dawn and dusk. Lions, hyenas, and jackals mostly hunt during the hours of darkness. Although wild dogs can kill animals up to 500 pounds in weight, they usually concentrate on prey weighing less than half that. Given a similar carnivorous pattern in the days of our ancestors, there would appear to have been an opportunity for a daytime carnivore who hunted prey bigger than that favored by the dogs. Inevitably there must have been clashes, both over the timing of hunting and the size of the prey, but there would still be a niche to be filled.

The animal bones excavated at hominid campsites almost two million years ago at Olduvai Gorge suggest that, at that time, our ancestors were concentrating on small animals, or the young of larger ones. But by one million years ago our ancestors were capturing big game in planned hunting expeditions. We can only guess how far and to what degree this goes back into our prehistory.

What, then, are we to infer about the lives of the early hominids from these observations of living carnivores? Basically, there are two things: first, the distinct advantage, in terms of biological economics, of hunting in groups; second, that group hunting implies at least some degree of cooperation, and that the more subtle the hunting strategies, the greater the cooperation demanded. To these may be added the social orientation that is a strong characteristic of the primate stock. What hunting must have done in the evolution of hominids is to re-emphasize the basic social pattern, weaving through it the vital thread of cooperation (not an important feature of non-human primates), and thus producing an animal with its own unique reasons for living in communities. Moreover, as we have stressed, our ancestors were not simply hominid carnivores: their economy was based on the special mixture of hunting *and* gathering, separate activities linked together as a powerful social pattern by the sharing of goods between members of the troop. Cooperation during a hunt could be, therefore, just one aspect of coordinated behavior by a group of social animals operating a mixed economy. Cooperation, tolerance, and independence were essential for this type of economy to be established initially, and more so for the subsequent sophisticated development.

Apart from the clear need among carnivores to range over extensive areas in search of prey, and form a certain intolerance of other hunting species, there appears to be no strong pattern of behavior that is inextricably tied to meat-eating in itself. Aggression and territoriality are both finely tuned to certain environmental conditions. Hunters are not subject to a steady buildup of nastiness inside themselves, such as must be periodically released either by making a kill or by starting a fight with some other member of the group. In other words, once again, hunting is not to be confused with aggression. Although various carnivores may appear to keep jealous guard over certain parts of a territory, this behavior is determined by the stress of population within the group, and by the amount of prey available. For instance, packs of wild dogs, which roam over hundreds of square miles, frequently share overlapping territories. The packs do not violate the areas containing their neighbors' dens, and there is no more than a general raised tension between packs if food is short.

One especially interesting fact about carnivores is the way modes of dominance vary from one species to another. For instance, packs of wild dogs and wolves, in which both males and females join in the hunt, have only a weak hierarchy, functioning only within each of the sexes (though the pack is usually led by an *alpha* male). Aggressive encounters in which individual members assert their status, or even attempt to climb the social ladder, are not a prominent part of their social behavior. By contrast, a pride of lions is often a tense social unit, in which the dominant male repels other males who have the effrontery to intrude, and also insists on first place in the meal queue even though the kill will more often have been made by the lionesses. Females have their turn among spotted hyenas, where they are dominant over the males.

Aggression, territoriality, and dominance can thus be seen as remarkably flexible features of carnivore behavior, and we cannot safely infer that any of these were typical among early hominids. Although the forces that nurtured the hunting evolution of hominids may have impressed a particularly human quality

on one of these behaviors, it is reasonable to suppose that inherent flexibility might have generated many different social patterns in different populations. However, the one tendency of hominid behavior that appears to have become very common, if not universal, is the social dominance of males over females, and this can be seen as a direct result of the division of labor between hunting and gathering. The current campaign for giving equal rights to women does its cause no good by denying this tendency. However unpalatable, its existence need not thwart the establishment of social justice today.

From our discussion of social carnivores it should not be surprising that humans and dogs have an intimate and long-established relationship, given the way of life the two groups have in common. The important difference between early hunting hominids (and their descendants) and the social carnivores, including the dogs, is quite simply that the carnivores depended almost one hundred per cent on meat for subsistence, whereas for contemporary human hunters, and almost certainly for our ancestors too, meat-eating was part of a nutritional and social package

The skeleton of a giant fossil elephant being excavated at a site about two hundred miles from Nairobi. Larger than present-day African elephants, its straight tusks reached the ground. Richard Leakey is crouching in the foreground.

which included large quantities of plant food: we were omnivores rather than true carnivores. Is it therefore valid to look to today's true carnivores for clues to our past? The answer must be yes, because, unlike most omnivores, the meat on our ancestors' menu came from big game. Omnivores generally obtain their meat by scrabbling around for anything from insects to lizards and rodents, not by organized hunts for big animals. There is therefore a crucial *behavioral* difference between the diet of typical omnivores and that of our omnivorous hominid ancestors. We should not forget, however, that whatever big-game hunting out ancestors indulged in was just part of a social organization which involved equally organized collection of plant foods: the social carnivores may give us *some* clues, but they cannot tell us everything.

The recent realization that we are not the only hunting primate – chimpanzees may hunt young baboons and other monkeys – and that baboons sometimes cooperatively chase infant Thomson's gazelles (though they have more success with spring hares) – does somewhat alter the perspective, it is true. But in both chimpanzees and baboons the absolute amount of meat eaten is small, probably accounting for between one and five per cent of their diet, as compared with possibly thirty per cent among the African hominids of two million years ago. They hunted rhinoceroses, antelopes and elephant-like creatures, as well as taking small prey and scavenging dead animals as the opportunities presented themselves. The mixture of meat to plants is less than around half and half, with meat probably being exceeded by plant foods.

From animal models we now turn to human ones: the few bands of hunting and gathering people that still inhabit marginal areas, from the baking Kalahari desert to the ice floes of the Arctic. These people cannot tell us how our ancestors made a living or occupied themselves during their leisure hours, but they *can* point to the possibilities and patterns of a hunting way of life. First we shall look at the general organization of hunting bands, and then in more detail at the day to day lives of hunting people.

One of the most intriguing discoveries to emerge from recent studies of hunter-gatherers concerns the structure of individual bands and the relations between them. Those bands are usually quite small, containing an average of twenty-five persons. About twenty such bands, or a total of five hundred people, generally make up what is known as a dialectical tribe – that is, a community whose members speak the same dialect. On the face of it, a band size of twenty-five may not seem particularly significant. Gorilla troops are usually just a little smaller, and so too are those of wild dogs; troops of baboons and chimpanzees are often not much larger. But the similarities probably obscure what is important about the human bands. In a group of twenty-five there will be seven or eight adult males. If a group of individuals have to engage in careful planning and execution of the activities called for by their mixed economy, then eight is just about an optimum number, as anyone who has sat through endless wrangles generated by committees of more than that number will know only too well. A band of twenty-five, containing about six families, is therefore probably close to the optimum for a cooperating unit.

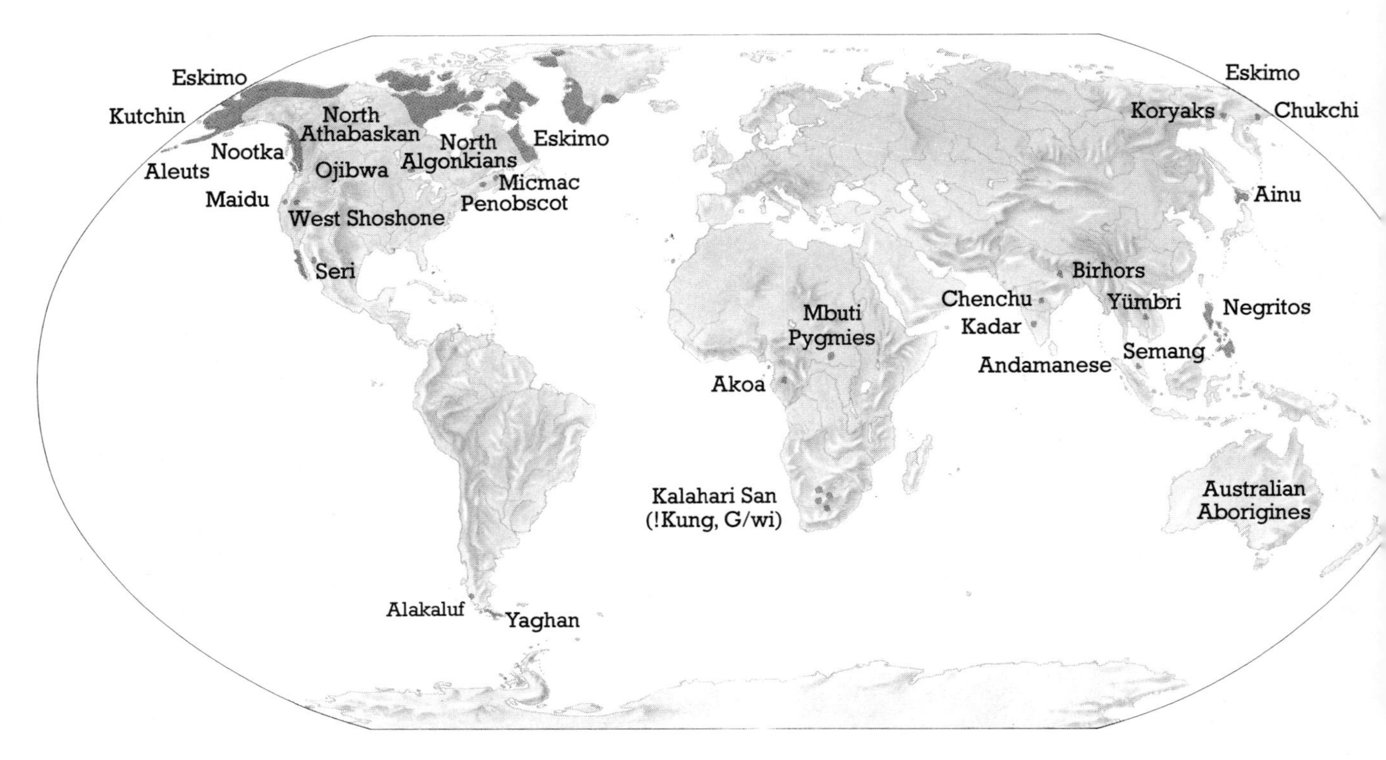

The number of hunter/gatherers today is a very small percentage of the present world population. The map shows the present distribution of the better documented hunter/gatherers.

Another very basic factor that must affect the size of bands is the availability of food. By its very nature, the large animal that constitutes potential prey must have a wide area in which to graze. Hunters therefore, must have access to an equally large area of land. And since they must have an adequate supply of meat to keep them alive throughout the year, they will be essentially nomadic, moving camp perhaps six times a year. Often, however, the move to a new camp is made because the local stock of plants is becoming diminished, and not by the lack of game.

The !Kung people of Botswana, for instance, occupy an average of slightly more than two square miles per person. This does not mean that as individuals they are not sociable. Quite the opposite. Their camps are unbelievably compact, with a great deal of unavoidable personal interplay. They evidently have no wish to avoid frequent contact, which is clearly an important part of their lives. And the amount of socializing between bands is astonishing: practically two-thirds of !Kung life is spent either visiting or being visited by friends and relations!

Although !Kung camps are in general widely dispersed during the wet season, during the summer drought they are set up in clusters around permanent water holes; during this time the men hunt in closely neighboring areas without any disruptive clashes over territory. Changes in climate or the availability of game may cause temporary modifications of the basic band unit of twenty-five. The G/wi people of the Central Kalahari, for example, live in an area that is even more marginal than the one inhabited by their neighbors the !Kung. With just a few inches of rain each year, they survive most of the time without any standing water. For perhaps six to eight weeks in total after the meager rains there are water holes, and it is around these that the band gathers. During the seemingly endless dry seasons the G/wi obtain water mainly by eating succulent plants. But these plants do not grow in sufficient density to support a band living in one place. So, when there is no water to drink the G/wi bands break up into units of three or four, usually comprising a nuclear family, which do what small-scale hunting and trapping they can – though most of their food comes from plants, and their water supply mostly from melons. The band reunites when the rains come.

So far as can be judged from the archeological remains, the hunting people in northern Europe some twelve thousand years ago practised a similar kind of band fission and fusion, dictated probably by the migrations of deer. These people, who were surviving through the tail-end of the last ice age, seem to have lived in groups of perhaps six families during the summer, when they could follow the herds of deer on their migrations. The abundance of meat from the deer was enough to sustain largish groups. But during the winter months, which must have been much longer than now, the disappearance of the deer meant that the hunters had to disperse to the forests where they lived as single family units subsisting on small game.

On the other hand, several bands may occasionally coalesce when game is briefly plentiful. An instance of this is found among the Bihors, who used to hunt monkeys on the Chota Nagpur plateau of Central India, and who exploited the annual herding of the sambur and axis deer to launch a concerted hunt. During the latter part of April and the beginning of May several bands (called tandas) would join together to prey on the herds. When these dispersed, so too did the tandas. One can imagine that our ancestors pursued similar strategies when opportunity arose. Indeed, there is persuasive evidence of this from Torralba, a site in Spain, where some three hundred thousand years ago, our ancestors gathered to hunt elephants, wild oxen, rhinoceroses, horses, and deer. (See Chapter 6, p.136).

A band of between twenty and thirty would thus seem to be at the heart of the social structure in a human hunting society. Because it was a tightly knit group, the individuals composing it would tend to be related to greater or lesser degrees, just as people are today who live in small, remote villages. A neat parochial arrangement like this is potentially dangerous, since unless there are amicable links between bands, clashes may arise between them. Moreover, a population of twenty-five is too small to be a stable one, not least because the balance between the birth of boys and girls is unlikely to be equal. Because in a hunting community where the men go after the game, such a balance is so essential, we can assume female infanticide to have been not uncommon among hunter-gatherers, to avoid having girls outnumber boys. This would not be such a problem among farmers, of course. A better means of arriving at a balance in hunting bands, however, would be to belong to a large group, the dialectical tribe. Assuming a band size of twenty-five, it is possible to work out the smallest tribal number that would ensure an overall balance of girl and boy babies, given reasonably constant birth rates and levels of infant mortality. That number turns out, conveniently, to be about five hundred!

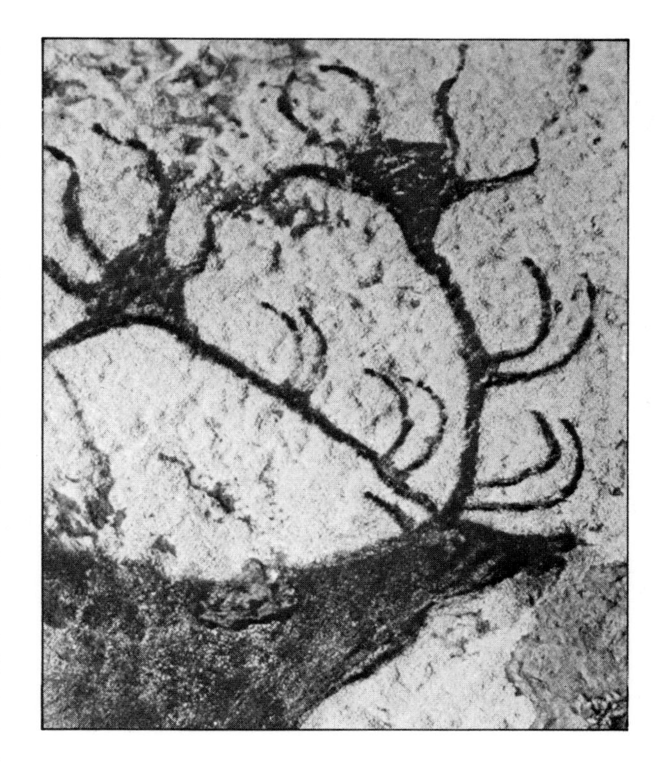

All the archeological evidence suggests that Cro-Magnon Man was a very proficient hunter. He understood the habits of large herd animals – the times of their migrations, the routes they used – and evolved efficient methods of hunting and killing them. It is probable that the migrations of deer (this cave painting is from Lascaux in the Dordogne region of France) dictated a variable way of life in terms of group activity.

Without exception, hunter-gatherers observe incest taboos, so that in general – most of the girls within the band being related to him in some degree – a youth must find his bride outside his own band. The practice of marriage between bands, or exogamy, is of course an excellent device for establishing links between neighboring groups of people. In addition to the incest taboo, many hunting people have further restrictions on who may marry whom. For instance, a !Kung youth must not marry a girl with the same name as his mother or his sister; and if a man is making a second marriage, he must choose a new wife with a name different from that of his daughter or his mother-in-law. Similar rules apply to girls. All of this means that some might have had to search farther afield to find a spouse – a process that would gradually extend the links of relationship across many neighboring bands. It is significant that, although !Kung hunters routinely go out on forays for food, by far the longest single journey a young man is ever likely to embark on is the one on which he finds his spouse.

A network of blood relationships will thus tie dispersed bands together over a very wide area, forming a cohesive unit whose components speak the same language, carry on the same culture, and have a broad sympathy for all members running through it. A second factor that may tend to limit the size of the tribe is the capacity for communication and memory. The dialectical tribe is informal in that relationships and cultural norms are in people's heads, and are not put down in writing. Very large numbers in a tribe would severely tax their collective ability to know and remember all they needed about one another to maintain a cohesive group. Whatever still undiscovered influences may have tended to nudge the tribal populations towards the five hundred mark, we can be sure that the tribe is a particularly human invention: it does not appear elsewhere in the biological world.

Traditionally, contemporary hunter-gatherers have been looked upon with a rather patronizing, if not altogether unsympathetic eye, as condemned to unabated toil in scratching out a meager living from the marginal areas of the world. According to the seventeenth-century philosopher, Thomas Hobbes, their lives were 'nasty, brutish, and short.' Now that anthropologists have begun to cast away Western pre-

Early morning in a temporary !Kung camp that has been established in a nut-tree grove.

Socializing is an important part of life in the temporary camps of the !Kung, as is shown by the activities illustrated here: telling a story (above); pegging out a skin (right); women playing melon-toss – a combination of a dance and a game (below).

The G/wi people live in an area that is even more marginal than the one inhabited by the !Kung. During the dry seasons they obtain water by eating succulent plants such as melons (left) and much of their food comes from plants such as the tubers (right) which are to be roasted. After a successful hunt the kill is carried back to the camp (above). It is interesting to compare the way in which a shelter is built by the G/wi (top) with the reconstruction of the shelter at Terra Amata on p. 129.

Life in a !Kung camp is not particularly arduous thus there is a fair amount of leisure time for activities other than hunting and gathering. On the left, a !Kung healing dance, and right, the !Kung camp, showing the shelters and the elements of 'affluence'. Strips of meat obtained from a hunt are hanging up to dry between the two trees on the right.

conceptions and to look closely at the way they actually live, a very different picture has emerged. In the words of the American anthropologist Marshall Sahlins, the hunting way of life is 'the original affluent society.' We shall see why.

It is to Richard Lee, who did one of the first thorough analyses of hunter-gatherers, as part of a long-term study of the !Kung organized by Harvard University, and his colleagues, that much of the current insight is due. Two main conclusions about such societies also shed some light on the evolution of our early ancestors. They are, first, that although hunting is indisputably a key part of social life, it is not the prime source of food; and second, that hunter-gatherers' lives are not nasty, nor brutish, nor short.

Today there are just a couple of hundred !Kung who still carry on this way of life in the northwestern area of the Kalahari Desert, a semi-arid region that wrestles with drought every two or three years. We now know that the !Kung have lived in much the same way here for at least ten thousand years, and probably longer – thus disposing of the notion that they have been driven into an extremely marginal region by the pressure of civilization. The !Kung mode of existence is long established, and clearly very successful.

The central focus of the !Kung's life is eight water holes which survive most droughts, and to which some dozen bands migrate during the dry season, lasting from May to October. The bands are, in fact, rather loosely organized camps, among whom there is a steady traffic of visitors. During this social season the !Kung spend approximately one-third of their time visiting other camps, one-third entertaining visitors and the remainder with members of their own particular band. Socializing is an important aspect of !Kung life. When the rains come the bands disperse to other water sources. But in either event, there is no problem over territoriality.*

* It is interesting to contrast with this the migration behavior of !Kung's neighbors, the G/wi people, who collect at water holes during the wet season (see Chapter 5 p. 109). No less than the !Kung, the G/wi are very keen on socializing, but for them it is the wet season that provides the opportunity. The socializing has many crucial roles, of course, including the arranging of marriages, so necessary to maintaining links between the dispersed peoples.

Gathering and trapping are a most important part of the !Kung economy; shown on the right is a kaross of plant food that has been gathered during a day's sortie and, on the left, a guinea fowl that has been trapped, together with a collection of eggs.

The diet of the !Kung is made up of roughly one-third meat and two-thirds plant foods. The !Kung have the particular luck to live where there is an abundance of the mongongo nut, a high-protein food. The !Kung eat about three hundred nuts each day, and this gives them 1,260 calories and 56 grams of protein – the equivalent of almost a pound of steak! The nuts, which are resistant to drought, constitute about a third of their diet, and thus amount to a secure and stable food source. As one of the !Kung remarked, 'Why should we plant [crops] when there are so many mongongo nuts in the world?' The rest of their plant food is a mixture of fruit, berries, melons, roots, and bulbs. The meat comes mainly from warthogs, kudu, duiker, steenbok, wildebeest, spring hare, guinea fowl, and various other mammals and birds, all of them rather small. Even more fascinating is the way the !Kung's food supply is collected. As in hunter-gatherer societies generally, it is the men who do the hunting. Occasionally they collect plant foods too, but only in passing them, and they report back to the women the location of good sources of fruits and berries noted during their hunting expeditions. These are undertaken with great enthusiasm but rather less success. The hunters usually go in pairs. Although there is great delight if meat is brought back to the camp, no one appears to mind too much if they return empty-handed, as frequently happens. There are always plenty of mongongo nuts!

Hunting among the !Kung turns out to be anything but a fulltime occupation. On average, men go out after game just two and a half days a week, and as each 'working day' is just six hours long, this amounts to nineteen hours a week – a total that falls somewhat short of arduous. Meanwhile the women are not hard pressed either. On one gathering trip they will generally collect enough to feed their families for three days, leaving each of the women plenty of time for visiting, entertaining, and needlework. There is rarely more than three days' stock of food stored in camp, which means that a steady pace of work and leisure is maintained throughout the year. An important aspect of leisure time is the so-called trance dance, an arduous ritual which is performed by the men and which usually goes on throughout the night. The participants rarely hunt on the following day.

The food economy of the !Kung is therefore heavily weighted on the side of plants and women. In terms of calories per hour, gathering is almost two and a half times more productive than hunting. Plant foods form twice as big a proportion of the diet as does meat. And still it is meat and the hunting to obtain it that are regarded with excitement and more than a little sense of magic. Although hunter-gatherers should probably more properly be called gatherer-hunters, there can be no doubt of the extreme social importance of hunting.

With a ready supply of food the !Kung maintain good health, and contrary to popular belief they do not die young, burnt out by pursuing an arduous livelihood under a baking sun. Roughly ten per cent of the !Kung are over sixty, a figure close to that in industrialized societies. The old are generally respected for their accumulated wisdom, both in practical affairs and in the niceties of ritual. And the young are not under pressure either. Youngsters do not contribute to the food economy until they are married, at around the age of twenty-three in men and eighteen in women. It is therefore the people between about twenty and sixty years old who hunt and gather; they constitute perhaps sixty per cent of the group. Childhood is carefree; adulthood is easygoing; and old age is relatively secure. Sahlins' definition of an affluent society as one 'in which all the peoples wants are easily satisfied' would certainly seem to be met by the !Kung.

Nor should one suppose that since the !Kung are favored by a virtually endless supply of mongongo nuts, their state of 'affluence' is somewhat atypical of the hunter-gatherers. The evident worldwide success of their way of life is entirely consistent with the notion that it was the mixed economy of hunting and gathering that brought our ancestors to the verge of agriculture. And if hunting and gathering were in fact a precarious mode of subsistence, human evolution would hardly have advanced at the rate it did up toward the revolution of ten thousand years ago.

The patterns of subsistence would have varied in different areas, of course. In the fifty or so hunting tribes that existed at least until recently, half have lived mainly by gathering, like the !Kung, one-third have concentrated on fishing; and the remainder mostly on hunting. Those belonging to the first group

The eskimos developed a meat economy that was based solely on the sea: the marine mammals provided not only food, but blubber for fuel, skin for clothing and bone for tools.

live mainly near to the equator; fishers in the cool and temperate climates, and those who concentrate on hunting in the Arctic. As one moves from the equator to the Arctic there is a steady reduction in the variety of edible plant foods: some Eskimos appear to exist solely on meat, eating no plants at all.

Since the cradle of human evolution appears to have been the tropical and subtropical regions of Africa, our ancestors are thus more than likely to have been hunter-gatherers, with the emphasis on gathering, and a crucial step in the evolution of *Homo sapiens* must have been the invention of the carrier bag. Without some form of receptacle in which to transport plant foods back to camp, a stable base for the food economy would not have been possible. The equipment needed for gathering is very simple: something like a stick for digging up roots and bulbs, and a container in which to transport the goods back to the camp, is all that is required. So, what is in effect a complex and particularly human behavioral pattern (that is food gathering within a society with a division of labor) is served by deceptively simple technology. The !Kung women use a kaross, part garment, part receptacle, in which they carry nuts, fruits, berries – and babies too. The kaross is made from antelope hide and is extremely efficient. We have to accept that even if our early ancestors had invented a kaross or something like it we would probably never know, simply because animal hide or plant material that might have been used in rudimentary weaving leaves a frustrating void in the archeological record. In the absence of solid evidence, one may nevertheless suppose that our ancestral hominids had more than probably invented some form of carrier bag by, say, three million years ago, and perhaps before that.

The carrier bag would thus suggest a key to the division of labor between men and women: the physically bigger, stronger, and faster men went out to hunt for meat, and the women provided the essential stability for the community, through the gathering of plant foods and caring and education of children. As with the !Kung and other contemporary hunter-gatherers, there was almost certainly more excitement about the men's contribution than the women's, even though the plant foods essentially kept everyone alive. There is a mystique about hunting: men pit their wits and skill against another animal, producing the silent tension of stalking, the burst of energy and adrenalin of the chase, and the elation of success at the kill. The challenge of a hunt is overt and visually impressive, the more so the bigger and fiercer the prey. Meanwhile the undoubted cerebral skills in mapping the distribution of plant foods, and knowing which will be ripe when, are much more calm, covert, and apparently unimpressive.

In the social web that held together the members of the ancestral hominid band, the reciprocal sharing of the spoils of hunting and gathering was the strongest filament. Emotional ties between parents (notably the mother) and offspring exist even in the nonhuman primates, particularly the gorillas and chimpanzees. To this basic structure, hunting and gathering adds a material dependence between males and females, thus tying the family knots even tighter. It is, however, more difficult to account for the family pattern of permanent monogamy to which Western societies are at least nominally committed. Most probably this pattern was born in the long-term material dependence through the accumulation of possessions that followed the agricultural and industrial revolutions.

The mongongo tree (right), where children are playing, provides shelter for campsites and nuts for food; on the left women dipping for water from the base of a tree.

Our hunting history is therefore long and it has left its impression deep within us. We can expect that as some of our ancestors ventured from Africa, a million years or so ago, into the harsher climate of Europe, they may have placed a steadily greater emphasis on meat. Throughout the transition from *Homo erectus* to *Homo sapiens*, which probably occurred around half a million years ago, and then to *Homo sapiens sapiens*, some fifty thousand years ago, hunting and gathering continued as the primary way of life, down to the invention of agriculture, ten thousand years ago. From the several parts of the world in which it sprang up simultaneously, it spread through the human population like a bush fire. Within eight thousand years at least half of the population had shifted to the new way of life. By two hundred years ago, hunters and gatherers had dwindled to perhaps ten per cent of the total population. During the first eight thousand years of agriculture the number of people on the planet rocketed thirty times, from ten million to three hundred million. And now, with the total population exceeding four thousand million, there are less than three hundred thousand hunting people left.

Right: The fertile crescent of the Middle East, where the agricultural revolution started. It did not, of course, happen suddenly – it was rather the culmination of many propitious facts: an accumulation of knowledge by the people living there of the plants and animals and their habits and environment, of the right climatic conditions and soils. The establishment of more settled communities and the storing of food that is implied probably led to fortifications – there was the need to protect what had been won with such difficulty – such as are seen in Old Jericho (above).

The speed of the transition is dramatic; no less so is the nature of the change. For the sake of our own future as a species, it is essential that we understand what has happened.

Hunters and gatherers, as we have shown, are very much part of the land that sustains them. To survive they must be in balance with what the land can offer them. They have a faith in the resources of their physical world, and in their own ability to exploit them.

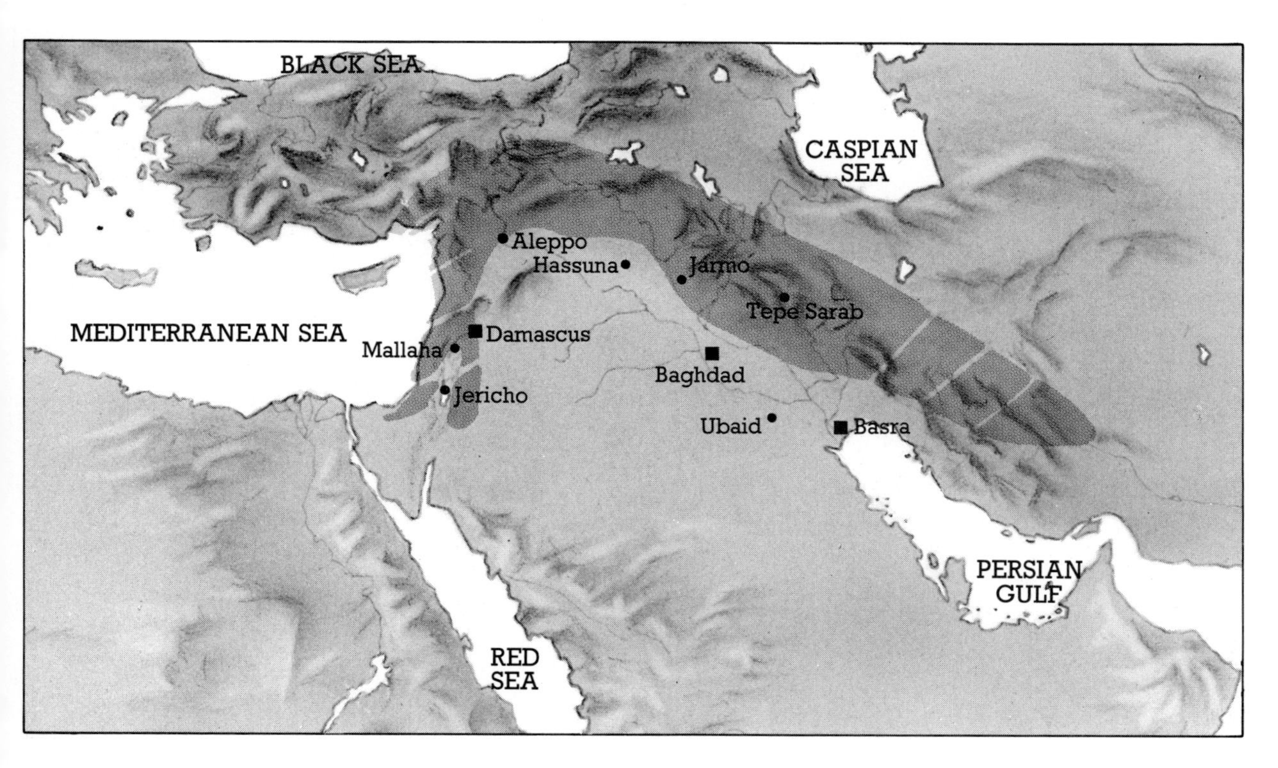

They live in small, intimate, cooperating bands as part of a diverse but familiar tribe, moving nomadically from camp to camp as food resources or their own whim may dictate. Hoarding material possessions is foreign to them, but this does not prevent them developing a rich culture and elaborate ritual. Hunters move over a land which they share with other bands and tribes not totally without confrontations, but certainly not in search of them as a necessity or a way of life. They carry their way of life wherever they go – less in the form of their few portable possessions, than in their intimate knowledge of their environment.

A farming culture is the exact opposite to this in almost every way. Because crops must be tended and the harvest waited for, farmers are obliged to be sedentary. And in turn a sedentary life offers for the first time the possibility of accumulating material possessions, calling forth a whole new aspect of human behavior which we may call psycho-materialism. The land bearing the crops must be defended, and so must the accumulated possessions. The farming economy can support a much greater density of people, thus giving rise to villages and towns – communities in which intimacy with everyone tends to become impossible as the population increases. With the growth of larger and larger populations, which become interdependent through trade, came the possibility of power over many people on a scale unknown to hunters and gatherers. And along with the possibility of expanding possessions and of power arose the concomitant urge to accumulate still more, combined with the need to protect what has already been won.

The technologically sophisticated world we live in today is torn by bloody clashes in the streets of urban ghettoes, and by the threat of military conflict between nations. Why? Writers such as the ethologists Konrad Lorenz and Niko Tinbergen, and the playwright Robert Ardrey, argue that these are the manifestations of genetically programmed behavior – in short that humans are innately aggressive. We believe this idea to be wrong. As we shall argue in more detail in a later chapter, so to account for the conflict in today's societies ignores not only the forces that made us human but also the dramatic social and psychological changes that followed the transition from hunting to farming.

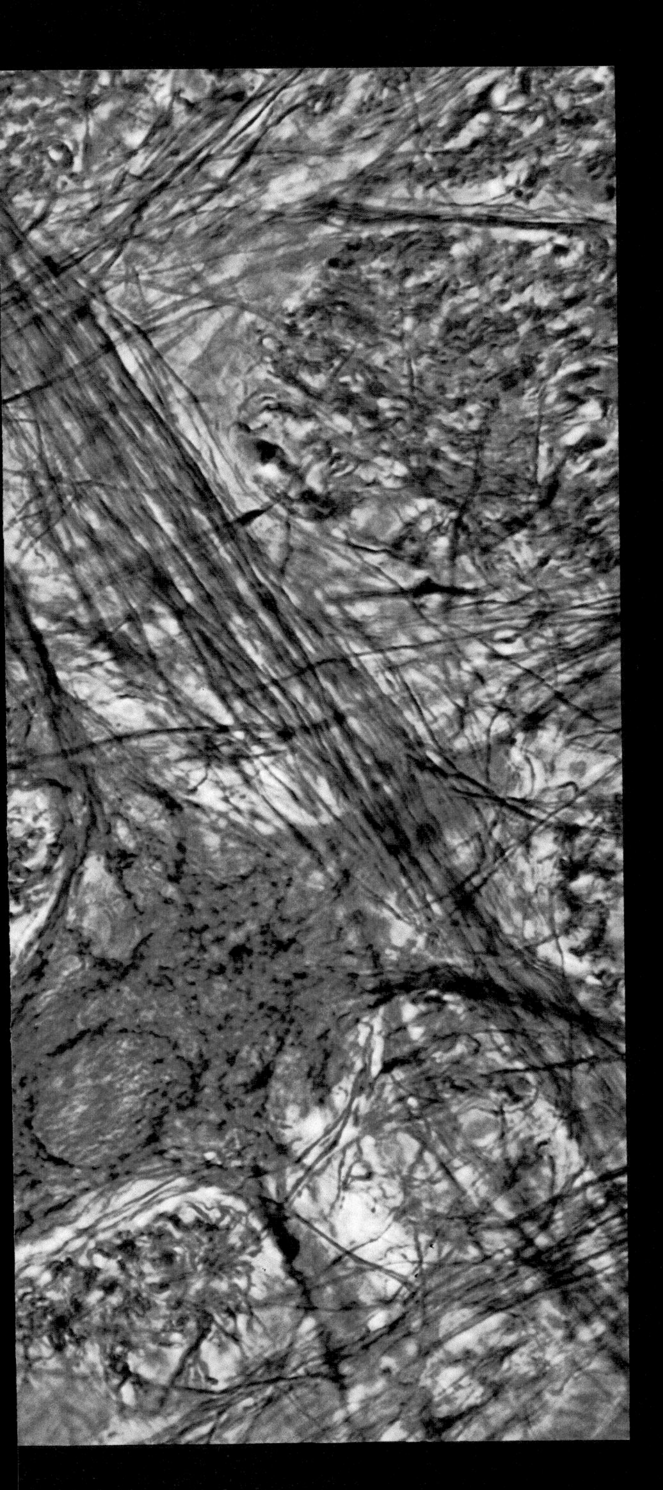

8 Intelligence, Language, and the Human Mind

When you talk to a very young baby, a curious thing is happening. Along with the arm-waving, the gurgling, and the engaging stare that can coax a smile from even the most solemn adult, the baby's body ripples with tiny coordinated muscular movements that can be detected only with specially sensitive electronic equipment. This extraordinary response, only recently discovered, is a powerful demonstration of just how firmly language is rooted in the human brain. In that baby the whole constellation of tiny, almost imperceptible movements have been generated by the brain as a result of the sound uttered by the admiring adult. More astonishing still – but also, on reflection, inevitable – is the discovery that exactly the same phenomenon occurs, whatever the permutation of nationality in baby and adult: that is, an American infant responds to the sounds of Chinese, Russian and French exactly as it does to those of the English language.

Spoken language is probably the latest step, and almost certainly the most significant one, in the evolution of the human brain. The ability to communicate verbally raises the possibilities of infant education to new and fertile levels, and is incomparable as a vehicle for the development and transmission of culture. Although it is possible to have a relatively rich cultural tradition in mute animals, through a consistent pattern of physical modification of the environment woven together perhaps by some form of gestural communication, spoken language as we know it magnifies immeasurably the potential subtleties of cultural organization.

A glance at the rich spectrum of the world's multitudinous cultures, which vary not only from country to country but also between neighboring villages, confirms the flexibility inherent in language. And, ironically, the increasing world domination of social patterns by a few particularly successful economic systems re-emphasizes the power of cultures as they swallow up and obliterate 'lesser' traditions that lie in the path of progress.

Previous pages: Man's brain is relatively large and very complex. Some idea of its complexity can be gauged from this photomicrograph of just two of the neutrons, or nerve cells, enlarged about one thousand times. The function of the neurons, which are in the cortex and of which there are several types, is to receive and transmit information. They communicate with one another and between them a total picture of experience is built up.

This example of a tablet with archaic text in semi-pictographic Sumerian signs is from Jamdat Nasr in West Asia – it is dated about 2800 BC.

The rich fabric of human cultures then, is woven from the words of a spoken language. This is what really separates humans from the rest of the animal kingdom (since we do not have to concede language to parrots and mynah birds just because they speak). But although the origin of language is so clearly a vital step, more than any other activity of our early ancestors, it is invisible in the archeological record. The only *direct* evidence of language is writing, and the earliest signs of it that we know appeared among Sumerian farmers, who about five thousand years ago recorded their stock holding on lumps of clay. It is safe to assume that our ancestors had command of a sophisticated verbal communication long before that. Just how long, we do not know – and if we are honest, we must admit that we probably never can. All we can do is to gather up the meager clues that may support an intelligent guess.

That is one of the aims of this chapter, and we link it in with a view of the origins – and meaning – of human intelligence. Language, tool making, and social organization become entwined in an evolutionary complex responsible for, and a consequence of, the

Many experiments to determine the level of intelligence and the manipulative skills of chimpanzees have been carried out. On the right Fifi is playing with a 'House of Cards'. Below, she is holding the handle of an automatic vending machine while she inserts a penny. As soon as the coin drops she pulls the handle to obtain some chocolate.

emergence of the special features of the human brain.

In the past, anthropocentric arrogance led to the notion that the first step in human evolution was the growth of a large brain – hence the eagerness with which the Piltdown skull was welcomed. According to a more recent opinion, the human brain was thought to be the result of recent and very rapid evolution. This also is probably wrong, as we shall presently explain.

A fundamental rule of biology is that an organism is adapted to the environment in which it lives. If there were such a thing as a totally stable environment – and the closest one comes to this is in the inhospitable depths of an ocean trench – an animal living there would be able to thrive with an equipment consisting of pre-programmed responses. Such responses are in fact a very useful attribute, so long as they are perfectly geared to their biological setting. For instance, a herring gull chick is innately programmed to peck at the red tip of an adult herring gull's beak – a form of behavior which in turn causes the adult to regurgitate food for the youngster. If a frail newly-hatched chick had to learn how to elicit that response from its parent, it might very well expire before it obtained a single beakful of food.

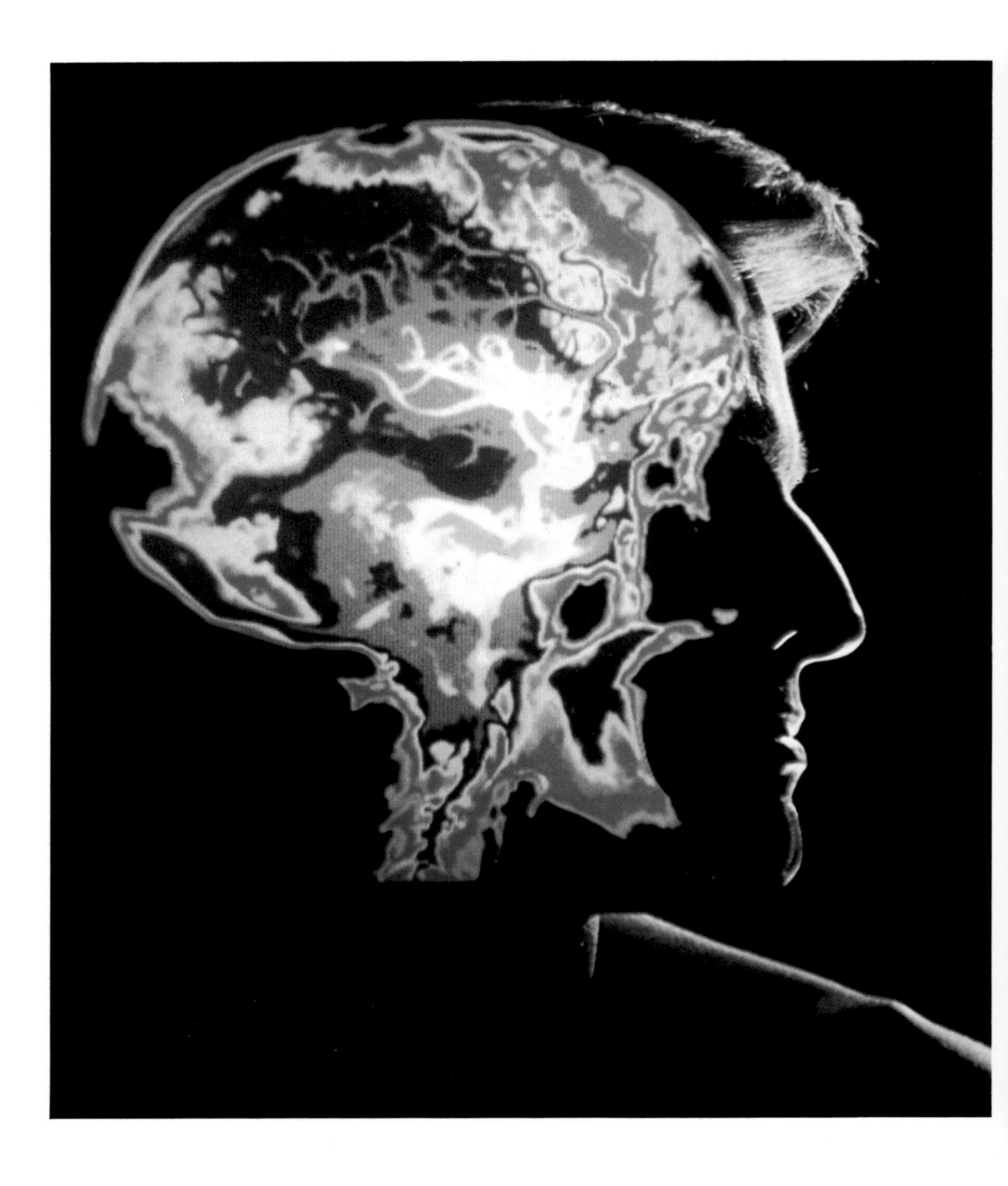

Scientific studies in many disciplines have, over the years, produced a basic picture of how the brain works. There is, however, much more to be learned. The various methods of brain scanning used for diagnostic medicine teach us more about that specialized structure which, in evolutionary terms, has pushed us forward to the unique position that we now occupy in the animal kingdom. The scan on the left is a color-coded photograph taken by an infra-red camera, while the other one (above) uses a radio-isotope (the patient has a serious tumor).

Although chimpanzees cannot talk, they do communicate vocally and non-vocally. Non-vocal communication is made by the general posture of the body, by touch and by facial expressions. A few examples of this last method are: normal relaxed expression (above), smiling (top right), pout face (below), and temper (below right).

On the other hand, an adult herring gull does learn to recognize its own chicks, and will unceremoniously expel any intruders – unlike its relative the kittiwake, whose nest is usually built on an inaccessible position on the face of a cliff ledge, where it is unlikely to be visited by other young chicks. As a result, an adult kittiwake will welcome any chick at all, of whatever species, that may be placed in its nest. Because of environmental necessity, the herring gull, which nests on the ground in populous colonies, must *learn* to recognize its offspring; there being no way in which an image of its future offspring can be pre-programmed into the adult gull's head.

Animals learn such things, then, as cannot be built into the innate structure of their brains and that will be advantageous to them in dealing with a variable environment. And in the economy of things, no animal is going to be equipped with facilities for learning more than it needs. Biologically speaking, the brain is a very expensive apparatus. In humans, for instance, the brain constitutes just two per cent of the weight of the body, and yet to function it demands fifteen per cent of the body's blood supply, and consumes more than twenty per cent of its total intake of oxygen.

Even in brains capable of learning, there are restrictions on when that learning can occur. For instance, young white-crowned sparrows learn to sing by imitation; unlike many birds, they have a song that is not totally controlled by their genes. There are, however, limits to their song-learning potential: a young white-crowned sparrow will not develop a song as melodious as a nightingale's, even if that is the only sound it hears. Moreover, it must learn its own particular dialect at a certain time during infancy, otherwise it will never be learned at all. The reason for this is clear. If adult birds were receptive to other dialects the whole function of group organization, of which song is a part, would be sabotaged. In much the same way, there is no doubt that the optimal time for humans to acquire a language is in early childhood. People who leave their native country to settle elsewhere usually retain strong traces of its accent for the rest of their lives.

The quality that accompanies the emergence of learning in the evolution of higher animals, namely intelligence, is surprisingly difficult to define. Basically, any animal that can modify its behavior by making use of information it receives from its environment may be thought of as intelligent. One of the qualities we have to explain in human evolution, however, is the development of creative intelligence, the ability to predict an outcome when confronted by a novel combination of events. Although many animals can predict on the basis of past experience, humans alone are accomplished at synthesizing evidence in quite this way. That is we can form concepts and manipulate them – in other words, we can think constructively.

But if we were to claim that the human animal is the only one equipped to perform this feat, we would be wrong. We have to admit that the great apes, particularly chimpanzees, appear to be much more intelligent than they have any need to be: the sophisticated challenges invented for them by experimental psychologists, to which they respond with such ingenuity, are at any rate far from what they are likely to meet in their natural habitat: this paradox poses problems for us in explaining the origins of human intelligence. But it may give us a clue as well; for whatever the differences, we do share one important trait with chimpanzees and gorillas especially, and that is in being highly social. It may be that the human facility for invention (the tangible manifestation of intelligence) is a by-product of the need to live in stable social groups. It may be, too, that spoken language is the by-product of another mental requirement: the need to think. But, as always in evolution, it is virtually impossible to distinguish the prime mover from the secondary consequence, and we should more properly view the development of intelligence as an integrated evolutionary process.

The early development of the human brain came as a consequence of an arboreal life: stereoscopic vision combined with grasping hands opened up for primates, particularly the higher primates, a three-dimensional world such as did not exist for other mammals. Objects within the environment acquired a significance in themselves because they could be picked up and explored: shape and color could be correlated with texture and weight, not forgetting that important mammalian form of inquiry, the sense of smell. The ability to act intelligently in the real world depends totally on the perception of that world. Yet the picture of the outside world you carry in your head is totally artificial. It is created by the mechanics of your brain, the information-collecting systems: eyes, ears, fingers, skin, nose – and memory.

The 'realness' of the world in your head depends both on the quality of the information collected and on

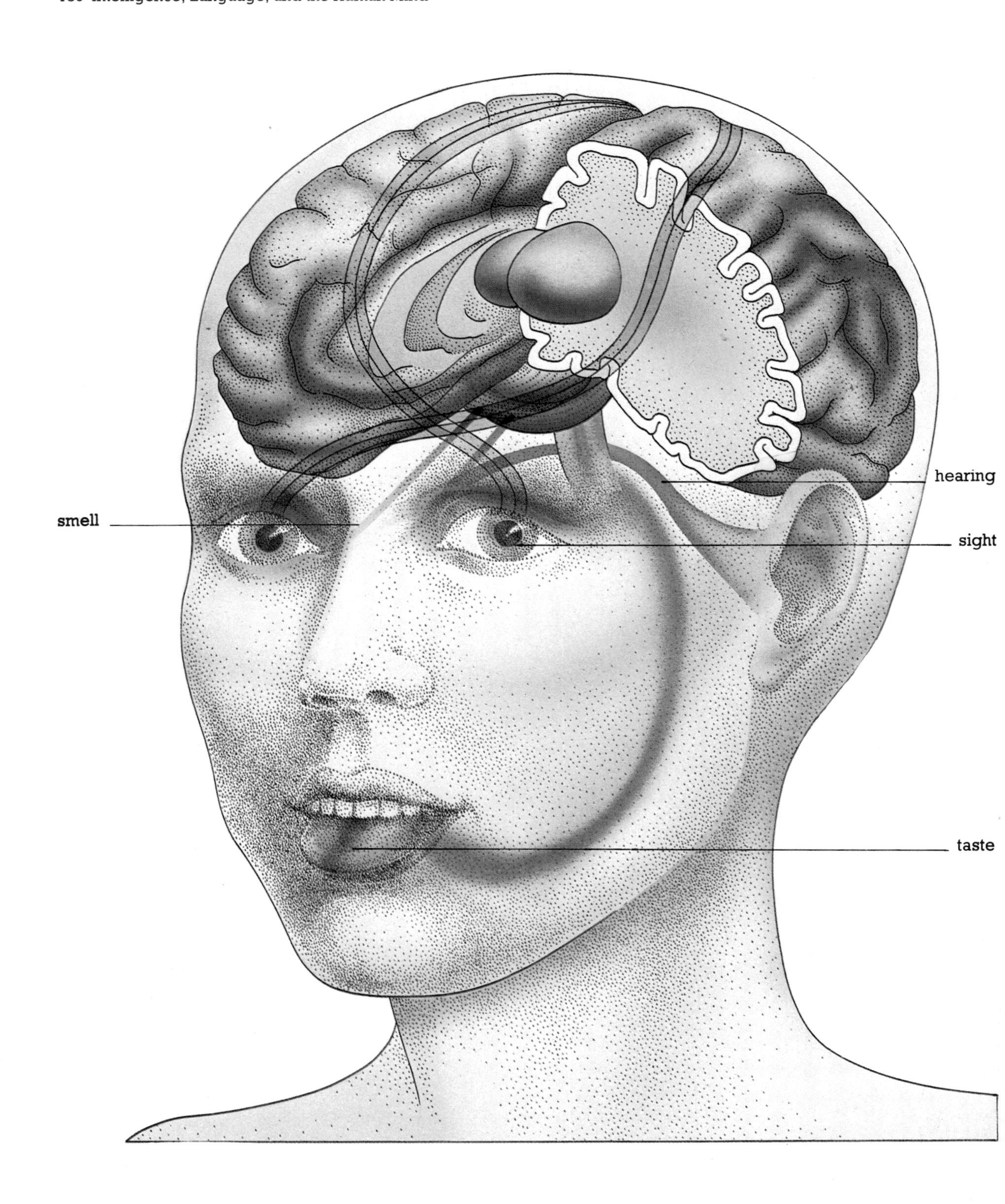
hearing
smell
sight
taste

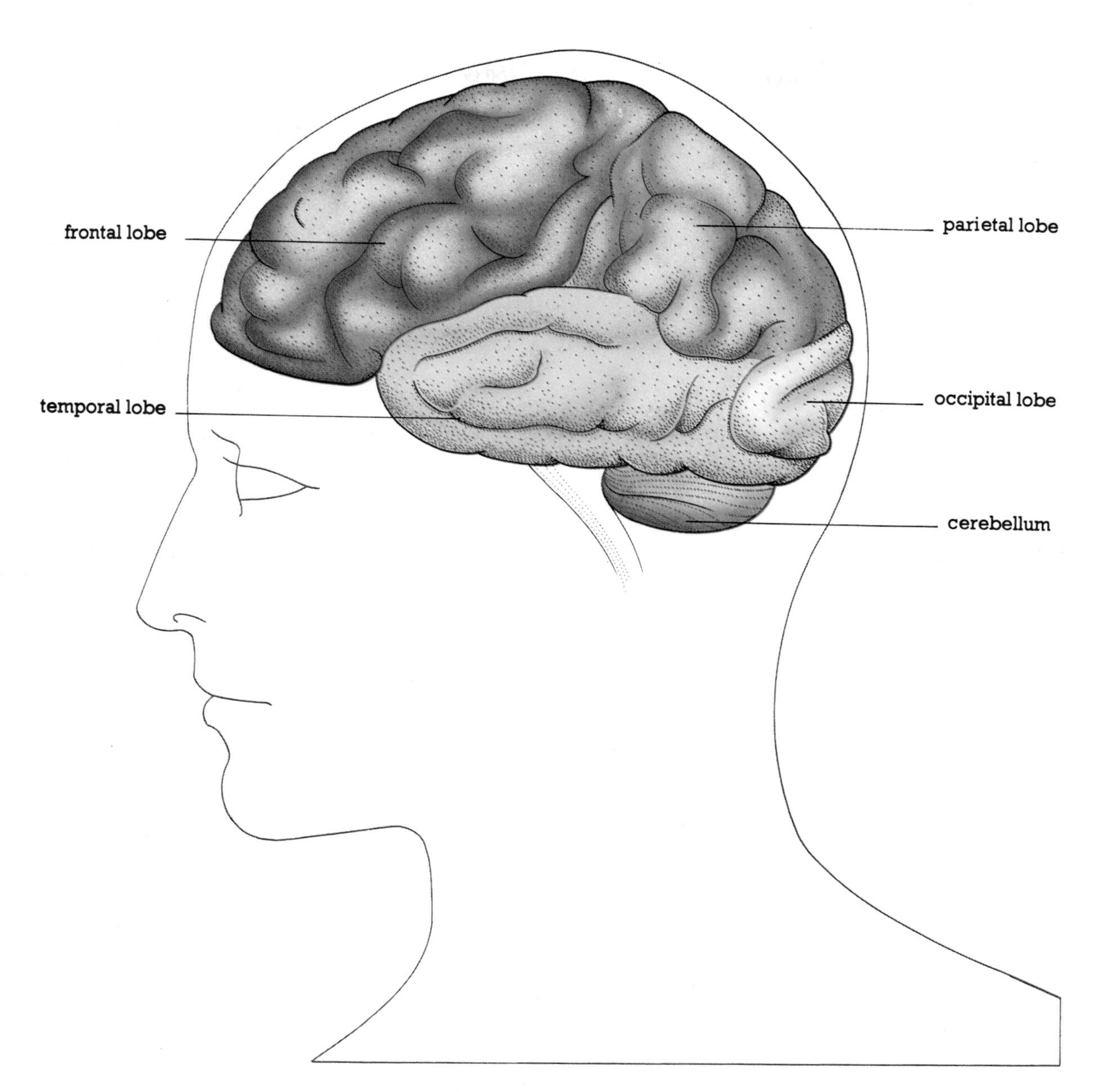

Left: The diagram shows the links between the brain and the sense organs in the head. The different pictures of the outside world provided by the tongue, the nose, the ears and the eyes are integrated in the cerebral cortex. Information provided by the sense of touch is added to achieve a complete picture.

Above: The cerebral cortex, which consists of convoluted greyish tissue divided into four lobes, is situated above the cerebellum – which is concerned with coordination and balance. The active nerve cells of the cortex (about eight billion of them) organize all the information that is fed into the brain from our various senses.

the way it is integrated into a coherent form. Images, whether visual or auditory, are generated and automatically held up for comparison with past experience. Events taking place now are interpreted in the context of times past, so that human consciousness not only exists for the present but also stretches back into the past and may be projected into the future.

The value of past experience is neatly demonstrated in an anecdote reported by Colin Turnbull, an American anthropologist who has studied the pygmies of the Congo region. The BaMbuti (the general name given to all pygmies in the Forest of Ituri) spend their entire lives so deeply surrounded by vegetation that the greatest distance they are likely to have experienced is no more than the few tens of yards between them and the other side of a river or a clearing. For the rest of the time their visual world is pressed in closely around them. It is against this background that they interpret the size and distance of objects they see. One day Turnbull took Kenge, one of the BaMbuti, with him on a long drive out of the forest and up a mountain overlooking Lake Albert. There Kenge, who found it almost impossible to believe in a world without trees, made a classic perceptual blunder. Pointing to a herd of buffalo grazing several miles away, he asked, 'What insects are those?' It took Turnbull a while to realize what Kenge was talking about. Because at that distance buffalo looked so small Kenge supposed they *were* small, in fact no bigger than insects. To a far greater extent than we realize, like Kenge we 'see' what through experience we have come to expect to see.

Throughout the evolution of perception, the trend has been towards monitoring more accurately what is happening in the real world and assembling it to form an ever clearer world, artificial but representative, in the brain. Our ancestors, ranging as they did over territories far wider than those exploited by their ape cousins, were obliged to construct a mental map of those territories, delineated from many visible reference points, so as to transmit the whereabouts of rich sources of fruit, nuts and other vegetable foods. This, in turn, would have placed an evolutionary premium on the capacity for detailed perceptual analysis of the environment.

A vital leap in the evolution of intellectual capacity would have been the ability to form concepts, to conceive of individual objects as belonging to distinct classes, and thus to do away with the otherwise almost intolerable burden of relating one experience to another. Concepts, moreover, can be manipulated, and this is the root of abstract thought and of invention. The formation of concepts is also a necessary, but apparently not sufficient, condition for the emergence of language.

Although we know that chimpanzees can form and manipulate concepts, in the wild their penchant for abstract invention is severely limited; they do not talk, nor, as far as one can tell, do they have a very sophisticated alternative channel of communication. And yet the intellectual gulf between humans and chimpanzees is not as wide as might at first appear. For instance, Julia, a chimpanzee raised by Bernard Rensch and J. Döhl, learned to accomplish a number of intellectually demanding tasks with an impressive degree of skill. In one of these, Julia had to work her way through a six-step series of locked boxes, each containing a key to open the next, to get at the last one, which had a banana in it as a reward. To hit on the right strategy Julia had to look through the transparent lid of each box to see what sort of key it contained, working back mentally from the goal box to the one containing the first key. As well as being adept with tiny keys and padlocks, screws and other novel challenges, Julia could guide metal objects through complex mazes with a skill comparable to a college undergraduate. Out of one hundred trials she got eighty-six correct. Rensch and Dohl commented that '. . . by watching the eye and head movements of the chimpanzee, we could in some cases state that the ape first looked to the exits (from the maze) and then to the path system near the starting point. Apparently she combined the sensations of the latter with the mental images of the goal region, a form of behavior which is very similar to that of man.'

In competition with biology students on the same maze test Julia on average was slower, but by no means every time. And Julia is no freak. Many psychologists have put chimpanzees through hoops of intellectual reasoning in which they display a degree of skill previously thought to be the preserve of the human species. Stretching chimpanzees' intellectual powers is a fascinating enough game in itself, but to discover why they should be so bright is still more intriguing, and probably more difficult. Clearly, the great apes in their natural habitats have little opportunity to flex their intellectual muscles in the way that experimental psychologists love to demonstrate. Indeed, the day-to-day lives of the great apes, particularly the gentle gorillas, are remarkably undemanding – at least, that is how it appears when one considers their basic practical task of subsistence.

Chimpanzees, for example, spend a surprisingly few hours each day collecting food, the rest of the time being occupied by a busy social life. Although gorillas appear to be considerably less sociable than their smaller cousins, they are nevertheless a tightly knit social group, with important ties between individuals. One of the reasons that gorillas and chimpanzees appear to have an easy life is that by the time they are adults they know a good deal about what is demanded of them: they know the lie of the land they live in; they know when and where particular types of food will be abundant; and they know what potential dangers to avoid. And the reason they know so much is that they spend a long childhood learning from their mother and other adults.

The link between prolonged learning and living in a stable social group is the key to intelligence in the great apes – and in humans. It is not that learning what has to be learned demands high intellect. It is the ability to manipulate the complex social interactions that are an enevitable part of group living that demands real quickness of mind. A chimpanzee troop spans at least three generations and may contain ten 'families' (each headed by an adult female). A young member of such a troop would interact mostly with his siblings, forming long-lasting ties in spite of the normal family squabbles. The security of family life provides a base from which to learn about individuals in other families, in relation to whom such things as status can be worked out. There is no doubt that just as humans are aware of individual temperaments and how they may be affected by particular social contexts, so too are chimpanzees and gorillas. And just as we choose 'the right moment', either to gain social advantage or to lend emotional support, so too do our ape cousins: one of the most intriguing discoveries to come out of the recent extensive field studies on chimpanzee troops is the complexity of social interactions and, even more surprising, the skill displayed by individuals in exploiting them.

It turns out that chimpanzees also trespass into yet another preserve, once believed to be exclusively human: the consciousness of self. It used to be said that many animals *know*, but only humans *know they know*. For chimpanzees at least, this is probably an injustice. To know in the 'human' sense is to be self-aware. In its extreme form self-awareness manifests itself in notions such as that of soul, but in simple form it merely means to be aware of oneself as an individual among others. The American psychologist Gordon Gallup demonstrated that chimpanzees are aware of themselves by the simple expedient of placing a chimpanzee in front of a mirror. It soon became clear, through a few judicious tests, that the animal recognized itself. And, this would be impossible unless the animal were truly conscious that such a thing as *self* exists. We can never actually know what is going on in a chimpanzee's head, nor indeed can any one of us know exactly what goes on in any other human's mind. But just as humans are able, through self-awareness, to put themselves in someone else's position, it is a fair guess that to some extent at least, chimpanzees can do the same thing with one another; and this is a very useful tactic in the strategy of social relations.

Intelligence is undoubtedly required for an animal to learn how to exploit the resources of its physical environment. But this does not match the skill demanded to operate successfully in the complex mercurial social milieu of group living: the location of a particularly fertile food source is soon placed in a mental map of the region; dealing with individuals whose reactions to the same event are variable depending on a myriad of circumstances is, however, much more demanding. So, it appears that the evolutionary process promotes its own progress: learning about the environment (which demands a certain intelligence) means living in a stable social millieu (which demands at least an equal and possibly a greater intelligence); as social intelligence increases, so too will the ability to learn; this in turn encourages an even longer social apprenticeship; and longer group living leads to more social intelligence. . . . This is not to suggest that social life was the prime mover in the evolution of human intelligence, but it would be difficult to argue that it did not play a very important role.

Chimpanzees' considerable reasoning skills do not greatly impinge on their world of practical affairs: even stripping twigs off leaves so as to fish for termites, and using leaves as a sponge to collect water, are skills readily learned from generation to generation. But the creative intelligence that a few million years ago enabled our ancestors to shape specific implements out of lumps of stone, and that today has opened up the possibility of interplanetary travel, is

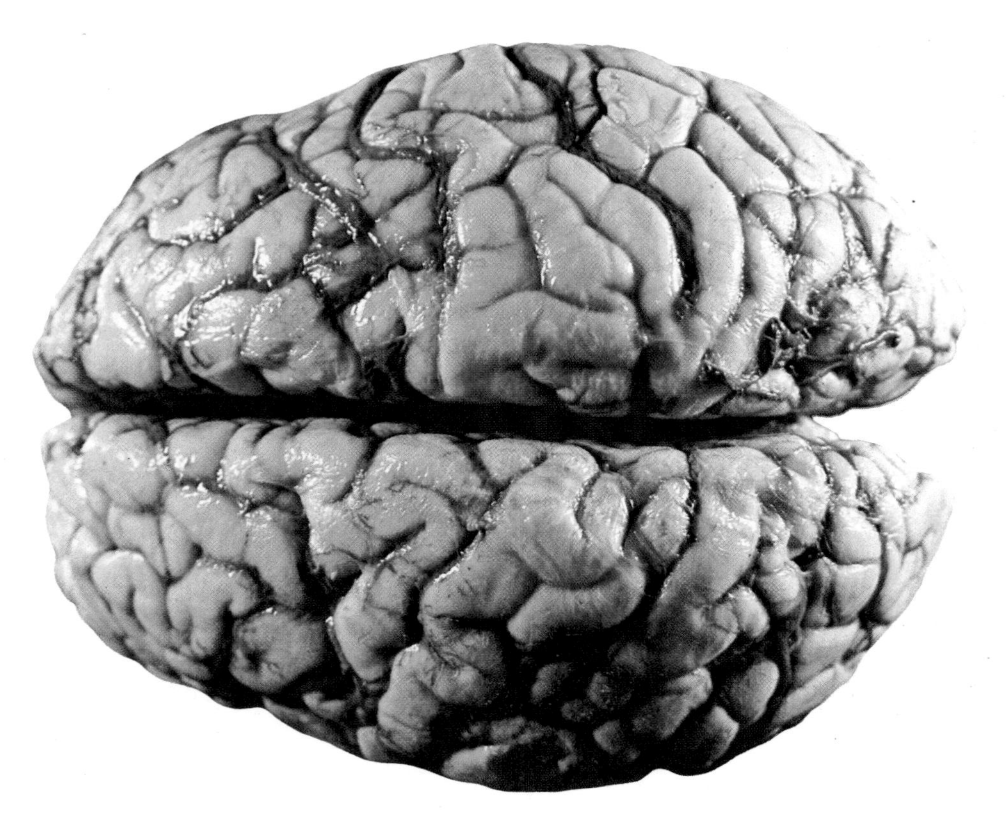

rooted in that same social intelligence, whose function is to mitigate the potential friction of group living. We *need* to be intelligent in order to thrive in groups; and we *use* our intelligence to be inventive.

As our ancient ancestor, *Ramapithecus*, moved from woodland to open savanna there was an inevitable tightening in its social organization; even in the early fossils we have seen the evidence of a prolonged infancy. As hunting and gathering and food-sharing also placed greater stress on group cooperation and organization, the hominids that thrived best were those that were able to restrain their immediate impulses and manipulate the impulse of others into cooperative efforts. They were the vanguard of the human race. Once tool technology had become an important aspect of hominid existence, that too would have operated toward selecting particular types of intellectual skills.

From looking at the tools themselves, we obtain only a glimpse of the society that made them. But the same tool that is used to slice tough roots may also be used to

The two halves of the brain are not identical as suggesled by the illustration above. The left half can be described as the 'logical' brain – it has those areas that deal with speech and the manipulation of numbers (see diagram) – while the right half, chiefly concerned with visio-spatial activities, is the 'intuitive' brain.

tear the skin from an animal killed in an organized hunt – which suggests how it might be possible for a whole way of life to change (from herbivorous to carnivorous) with no particular evolutionary pressure toward the invention of new tools.

The evolutionary trend that culminates in the human mind has been a simple well ordered and biologically economical one. Just as the human body has escaped the shackles of extreme specialization, so too the human brain has succeeded because of its flexibility. The secret of the human mind is that rather than having the ability to learn variants of *specific* tasks or behavior patterns, it simply has the *ability to* learn, to be

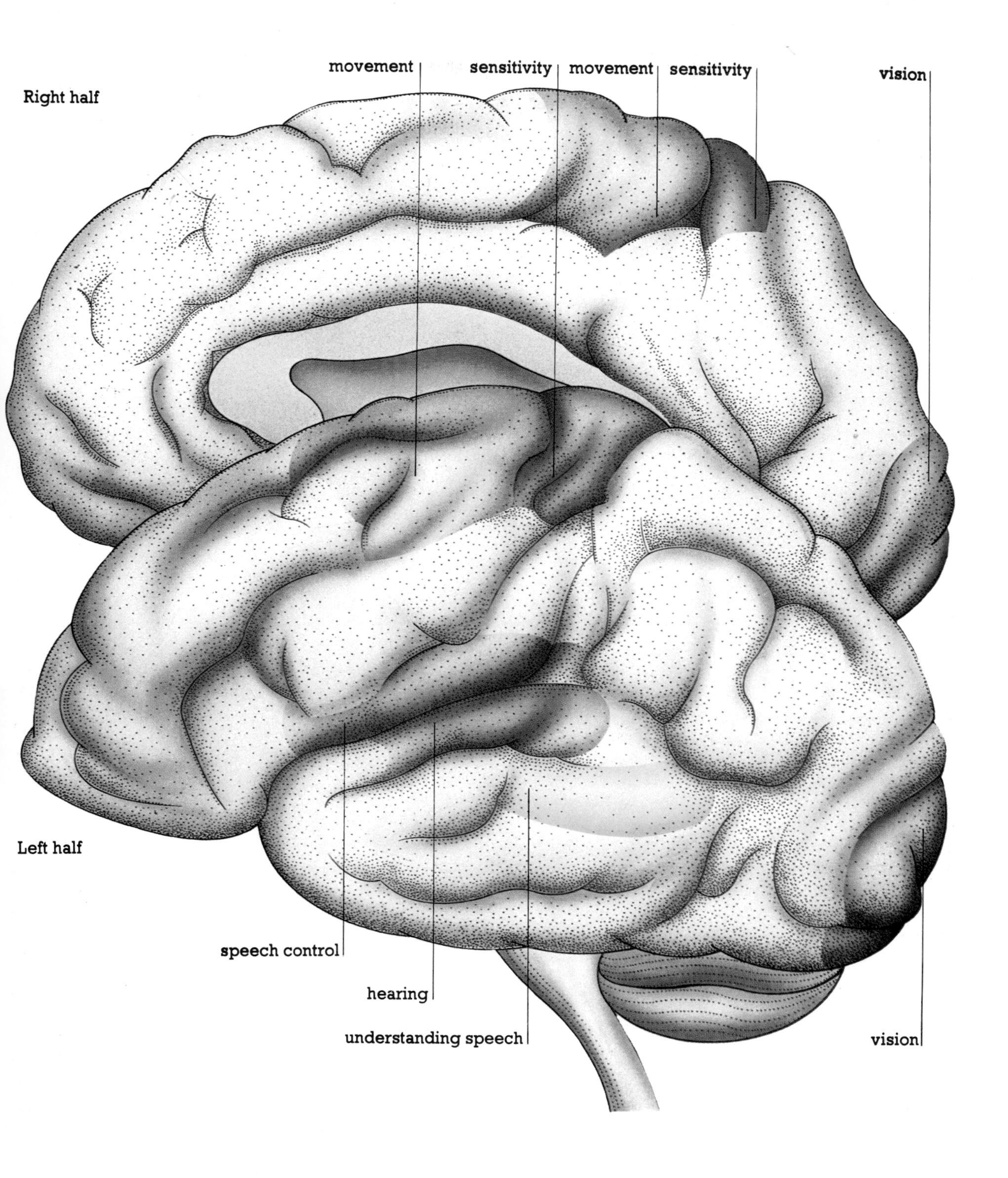
movement
sensitivity
movement
sensitivity
vision
Right half
Left half
speech control
hearing
understanding speech
vision

adaptive to practically anything that the environment has to offer.

Animals need to gather information about what is going on outside their heads so that they can construct some kind of representation of it inside. An animal's world is only as real as the information that channels into its brain. The more information the brain has, the more real will be the reconstructed world. But the signals from the ears, nose and eyes do not remain separate; they are integrated to form a more complete picture, and this integration is performed by the outermost crust of the brain, the cerebral cortex. It is this part of the brain that shows the most dramatic structural advances through evolution, and in the human brain it becomes the crown of biological success.

As the cerebral cortex expanded to carry out its integrative functions on which more and more complex behavior patterns could be based, it covered the whole of the more 'primitive' brain, and was finally forced to gather itself into folds to increase the area still further. Thus, although the chimpanzee cortex looks superficially similar to the human's, a mere twenty-five per cent of its area is buried in folds, whereas that of the human brain is convoluted to a total of sixty-five per cent.

The brain is split down the middle into apparently equal left and right sides. In humans at least, the equality is only apparent; the two sides have in fact become specialized for different major functions. In most people the left brain is more 'logical' – it deals with speech, word memory, and analytical tasks; the right brain is more intuitive, more 'artistic' – sited there are spatial abilities and skills for relating accurately to a three-dimensional world. This kind of division of labor in the brain is probably to some degree the result of evolutionary pressure toward a more efficient use of brain machinery: it increases the organ's working capacity without increasing its size. This specialized division of the brain power had until very recently been held to be unique to humans, but there are increasing hints that at least the elements of lateral specialization are already present in Old World monkeys. Even if brain lateralization, as the trait is called, is not unique as such to humans, the degree to which it has been specialized certainly is.

Some psychologists say that men are better than women at skills demanding spatial abilities, whereas women are far more accomplished verbally. This fits in with the probable division of labor in our hunter-gatherer ancestors: the males, as hunters covering large tracts of land in search of their prey, would have been given an advantage by the trait of spatial perceptions; on the other hand, females, who spent more of the time in a sociable camp atmosphere, engaged in full-time education of the youngsters, would have had a greater need for verbal skills.

The deeply folded cerebral cortex is divided into four areas or lobes. The frontal ones, the expansion of which forced the formation of the high forehead so characteristic of humans, are somewhat mysterious; all we can really say is that they have something to do with initiative, motivation, and restraint. In general, non-human primates are rather poor at concentrating for a long time on a single task; they lack persistence. For an animal to be a successful hunter, one quality it must have is persistence. A carnivore that cannot keep its mind on the job soon goes hungry; unlike plants, food on four feet usually moves on. The prominence of the frontal lobes may therefore be in part a product of our hunting past.

Also lodged in the frontal lobes, or at least the one on the left side, is Broca's area, a part that is concerned with the structure of speech (the grammar) and with initiating the mechanics of muscle movements in the face, lips, tongue, and larynx. One simple reason why apes cannot talk, in spite of intensive tuition, is that they have only a poor Broca's area.

Broca's area is linked by a bundle of nerve fibers to what is called Wernicke's area, located in the temporal lobe, the major storehouse for visual, auditory, and verbal memory; from this storehouse Wernicke's area plucks out appropriate words, to be assembled in their turn into structured speech by Broca's area. It is no accident that this second speech area is close to one of the most important developments in the human brain: the so-called superassociation area. A feature of an advanced brain is the way it squeezes every sliver of information out of the signals it receives, through bringing together and comparing these in what are known as association areas.

This technique of comparing and integrating signals is raised to a peak of achievement in a veritable orgy of information-swapping in the superassociation area: messages from the eyes, ears, nose, and skin feed in here so that, for instance, when you pick up a cup you can combine the sensation of shape with what it looks like; and if someone says. 'That's a cup,' you have a

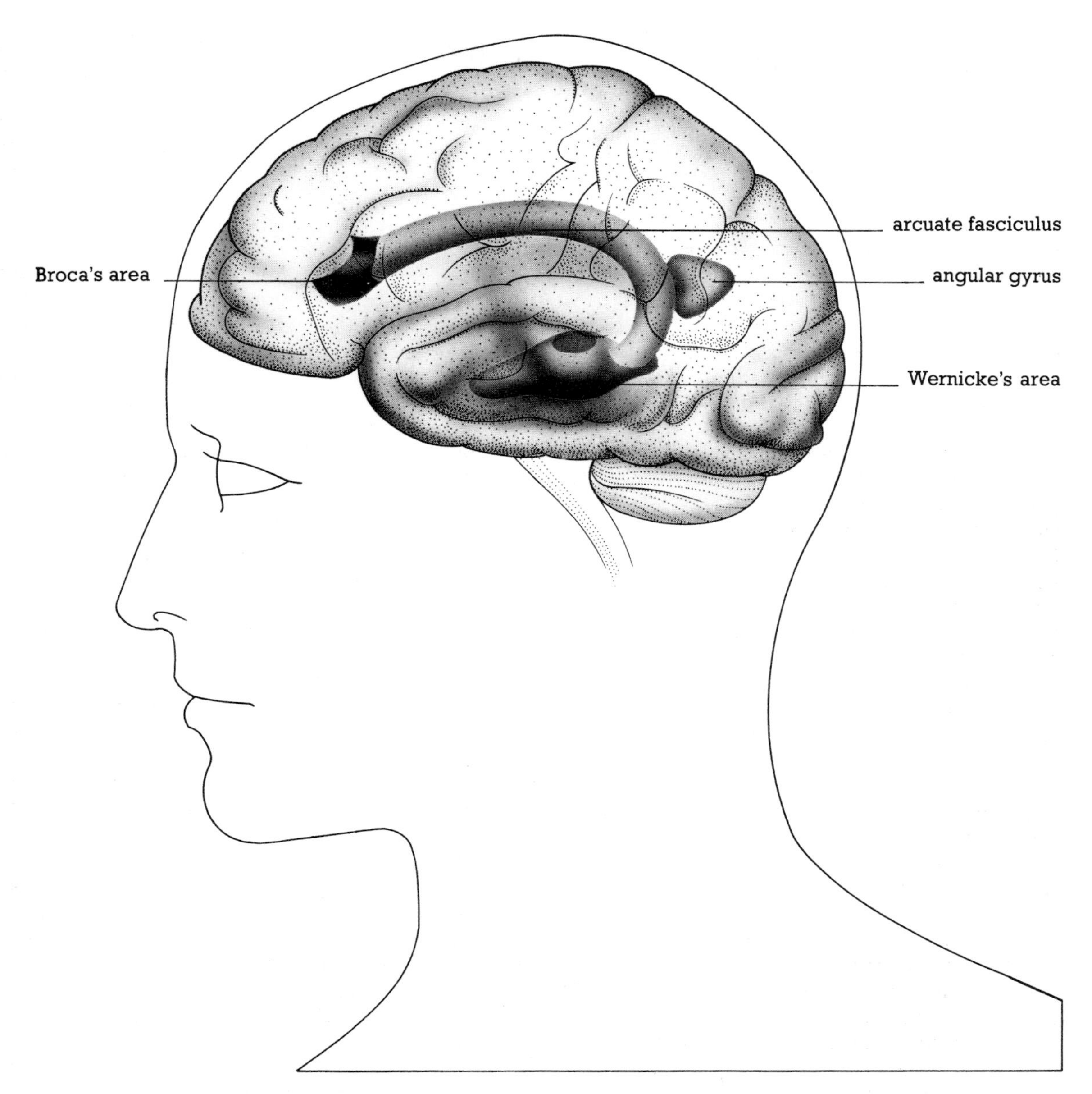

label for it too; and the label can be written *and* spoken. Without the integrative skills of the super-association area, a complex language would be very difficult to handle. The fact that monkeys and apes have a relatively small association area as compared with humans means that their construction of a mental world is correspondingly inferior; so, although chim-

On the left side of the brain in the region dealing with speech are three important areas: the angular gyrus, which integrates messages from our senses of sight, hearing and touch; Broca's area, which transmits instructions to the speech mechanism; and Wernicke's area, which deals with the process of comprehension.

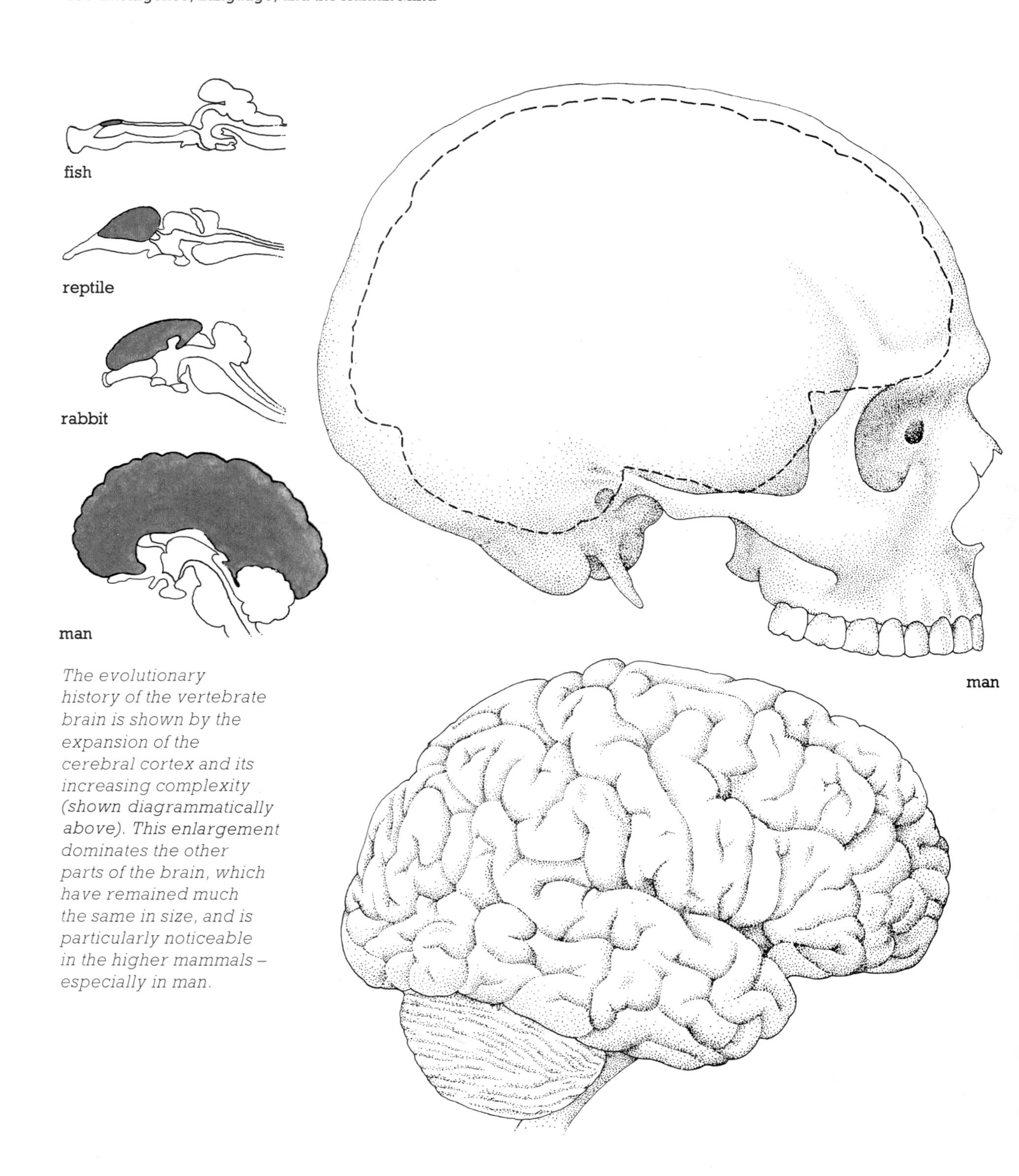

The evolutionary history of the vertebrate brain is shown by the expansion of the cerebral cortex and its increasing complexity (shown diagrammatically above). This enlargement dominates the other parts of the brain, which have remained much the same in size, and is particularly noticeable in the higher mammals – especially in man.

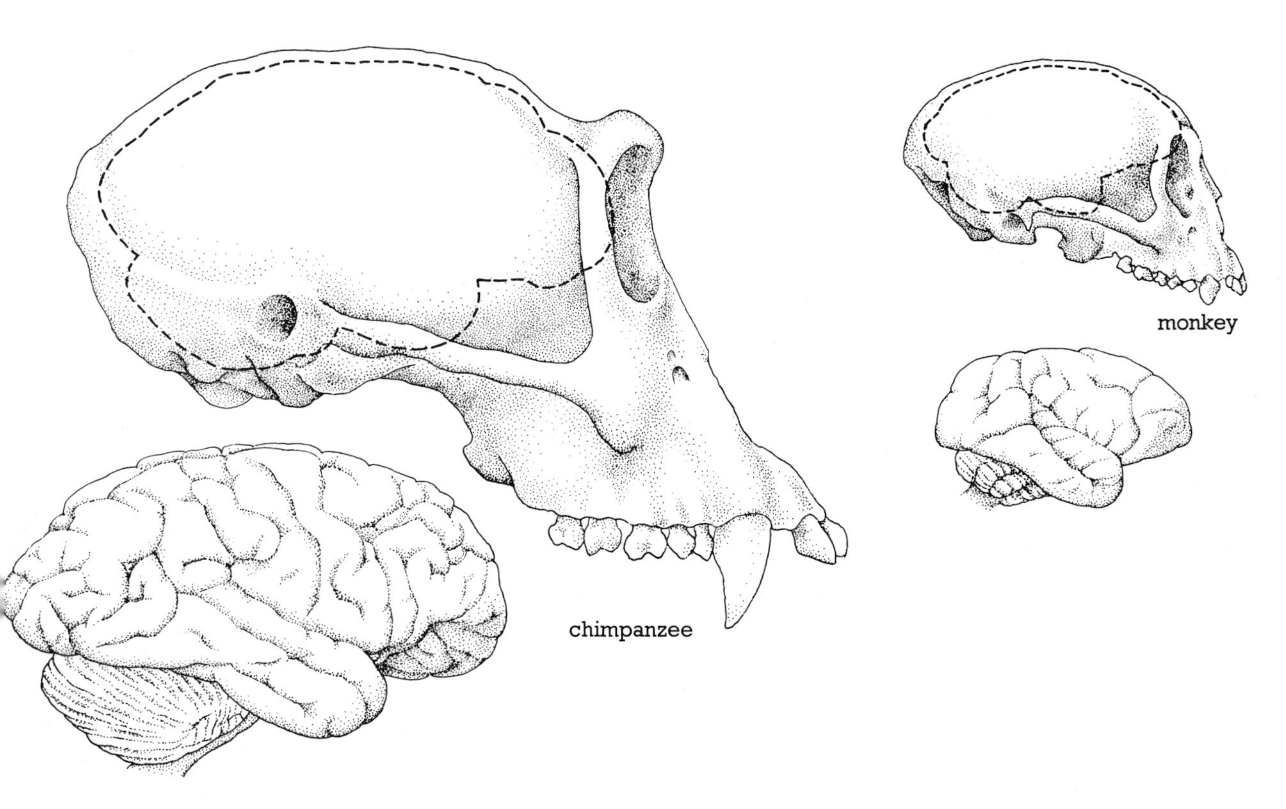

Not only is the human brain much larger than that of a monkey or a chimpanzee, but it also has a far larger area of convoluted cortex. The outline in the skulls shows the comparative sizes of the brains, while the drawings of the cerebral cortex (not to scale) illustrate the increasing convolutions as the evolutionary path is ascended.
The depth of the convolutions of the cortex is shown in this stained cross section of a human brain.

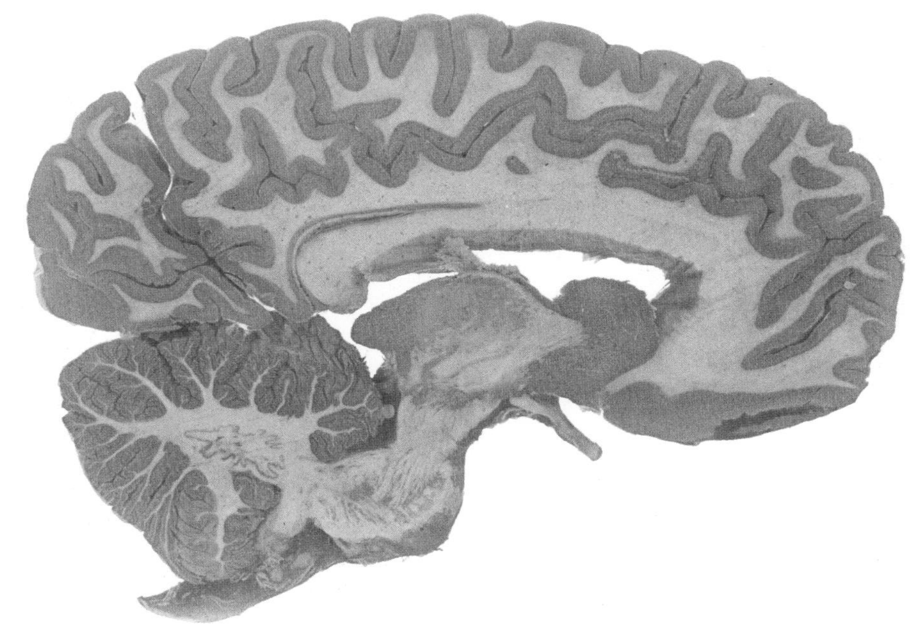

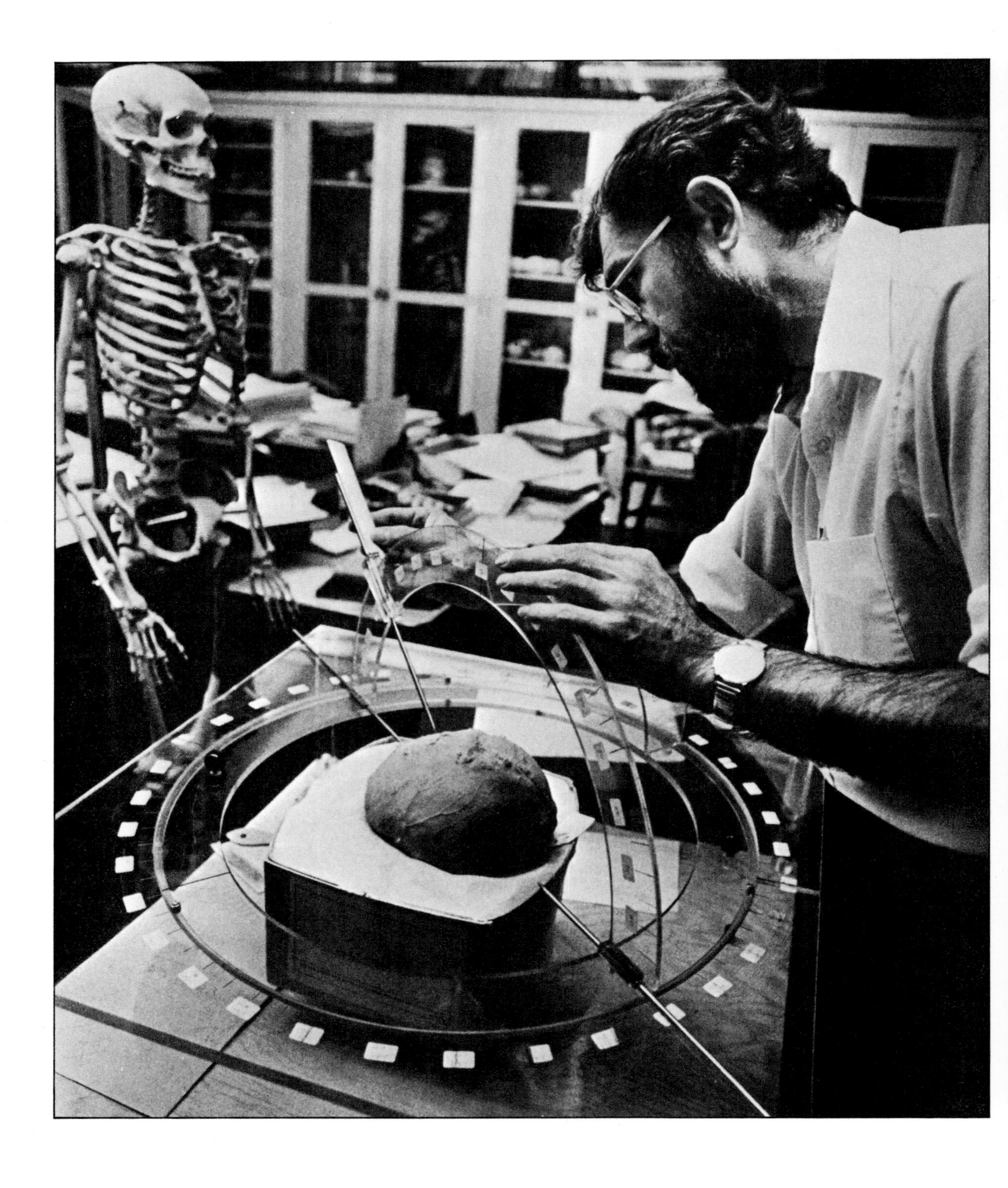

panzees and gorillas are capable to some degree of sign language, that degree must always remain limited.

The parietal cortex, which houses part of the superassociation area, is principally involved in assessing information from the sensory channels and in organizing appropriate responses. The expansion of the parietal and temporal lobes in humans has forced much of the occipital lobe, which deals with messages from the eyes, to become buried, making it appear relatively small.

Forgetting for a minute their relative size, one can distinguish between a chimpanzee's brain and a human's by the shape of the cerebral jigsaw: basically, a brain with small temporal and parietal lobes and relatively large occipital lobes is apelike; by contrast, a brain with a human organization has large temporal and parietal lobes, and relatively smaller occipital lobes. We stress the organization rather than the size of the brain because this is precisely what is most important in assessing its basic 'humanness.' There are, for instance, many dwarfs with brains hardly larger than a chimpanzee's, and, although they are in many ways mentally retarded, they can often speak fluently, an essentially human attribute.

The size of the brain is obviously important in an absolute sense, however, because there must be a limit to how much a particular piece of brain tissue can do and how many memories it can store. But the importance of brain size has undoubtedly been exaggerated in the search for the evolution of 'humanness,' while the much more crucial factors of organization have been neglected. The reason is obvious enough: until recently it had been impossible to get any good idea of the external structure of ancient brains, since their lobes do not leave a detailed imprint on the inside of the skull. But that interior is not entirely featureless, and it is possible to reproduce a 'shadow' of ancient brains by taking a latex mold inside the brain case (to produce an endocast), a technique developed by the American biologist Ralph Holloway. With a reasonably well-preserved fossil skull it is possible to arrive at some idea of the general proportions of an ancient brain and to determine whether it was apelike or humanlike.

Ralph Holloway has developed a technique for reproducing a 'shadow' of early man's brains by taking a latex mold of the inside of a brain case. Using his measuring device he is able to compare the dimensions of the molds with those of a replica of a modern brain.

It comes as something of a surprise to discover that *all* the hominid skulls Holloway has examined so far, including representatives of *Homo* and *Australopithecus* going back as much as three million years, apparently housed brains with cerebral jigsaws more humanlike than apelike. Thus, whatever evolutionary pressures molded the human brain had been operating long enough to shape the basic pattern by at least three million years ago. By two million years ago, divergence in gross brain organization is apparent between these two members of the human family: in *Homo*, the temporal and frontal lobes are becoming enlarged, indicating the powerful evolutionary forces at work during this period, steadily enhancing the humanness of our ancestral brains.

Even so, the notion that the 'humanness' of the human brain is the result of recent and very rapid evolution is probably not valid. And the notion is further undermined if one takes into account the smaller body sizes of our ancestors and their relatives. Although the size of the modern human brain is impressive, it is certainly not outstanding in relation to the size of the body which it serves: among mammals, the squirrel monkey, the porpoise, the house mouse, and the tree shrew all score higher on a scale of brain weight/body weight. However, among the hominoids (that is, the apes and the human family) the brain of *Homo sapiens* comes out on top. But what we are most interested in is a measure of braininess *within* the human family, extant and extinct: is the apparent acceleration of human evolution over the past four million years, and particularly the last one million, matched by a dramatic expansion in brain power relative to the bulk of the body?

This is an important question, not least because a brain size of approximately 700 cc is often viewed as the threshold of humanness: a hominid that had passed this hurdle could be accorded the status of *Homo*. This is clearly a somewhat arbitrary measure, but unfortunately it is about the only one we can be sure of: whereas we must admit that since the capacity of fossil brains is easy to obtain, the precise brain size that constitutes a 'ticket to humanness' is not.

An attempt at a more sophisticated analysis has been developed by Heinz Stephan and his colleagues at the Max Planck Institute for Brain Research in Frankfurt. In essence this technique scores the size of an animal's brain in relation to a standard animal, such as the tree shrew, which is very like the basic stock from which all primates radiated. The more 'brainy' an animal is, as compared with the standard animal, the higher its score (or progression index as it is called). For modern humans the average score is just under 29, though the spread is very wide – from a low of 19 to a maximum of 53.

Our closest living cousin, the chimpanzee, has a brain about one third of the human size, and with that it scores 12 on the index. By contrast, the small *Australopithecus*, with a brain roughly the same size as the chimpanzee, but with an estimated body weight of around forty pounds, comes out with a score of just over 21. This is respectably within the human range. We must stress once again that the fossil evidence for early hominids is less than superabundant, especially for so sophisticated a study. Nevertheless, all the hominid brains from the two to three million year period that have been put through the hoops of Stephan's progression index emerge with scores sometimes found in modern human populations.

There is just a hint of a slight 'improvement' in relative brain size through the ages in Holloway's finding that for specimens of *Homo erectus* the index is marginally short of 27. Until many more skulls are tested, the idea of a firm average must remain elusive. And we can still be less than firm about precise body sizes too because the evidence from fossils is simply too scanty to allow us to build up a clear picture. What is very clear, though, is that if there is some evolution in the relative size of the brain over the past three million years, it is probably nothing like the dramatic explosion that has frequently been imagined.

That the brains of our ancestors, and their close relatives, some three million years ago were equipped with a *basic* hominid organization would seem to argue that the 'humanness' of the human brain is rooted deep in evolutionary history, possibly long before tool-making and serious hunting were part of the hominids' everday life. What shaped the human brain from its apelike origins would seem to have been a matter of the social responses to the new ecological niche that was being carved out by our ancestors: a tighter, more complex social grouping, bringing with it a greater need to manipulate the ramifying relationships, to behave cooperatively, and to be more aware of the surrounding terrain would all have exerted keen selective pressures. These pressures would later have been sharpened by the social and mental demands of more and more sophisticated hunting and, to a much smaller extent, by the invention and development of tool technology.

Tools are an unreliable guide, however. The staggering advances in practical technology that the world has witnessed during the past thousand years, or even the past two hundred, are not in themselves the results of increasingly powerful brains. Rather it is the products of the brain – knowledge and culture – that have advanced. These advances tend to take place at an ever accelerating rate because of accumulated knowledge and practical skills. It is therefore probable that in the early stages of evolution when the store of practical knowledge was still modest, advances in social organization and sophistication would

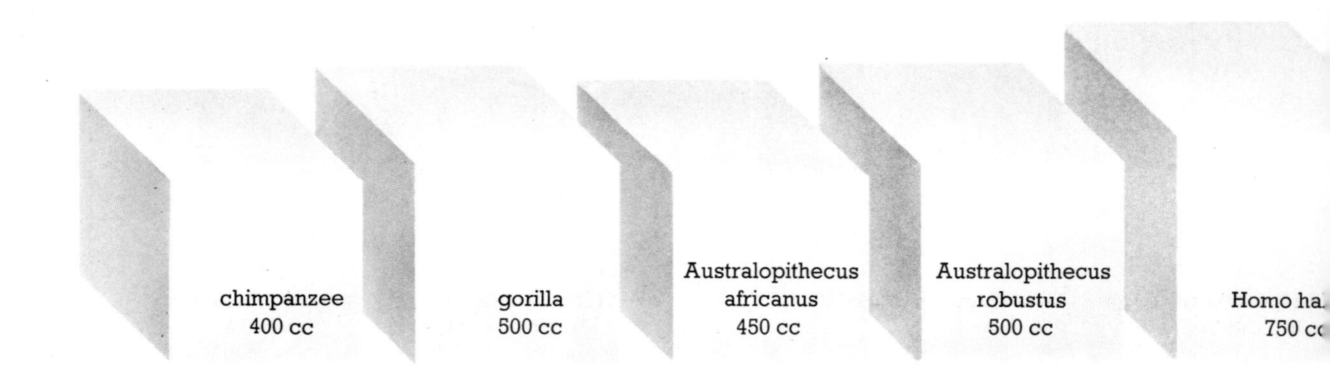

have been concealed behind an apparently static tool technology

As the size of evolving hominids' bodies increased, there was a natural demand for more brain power simply to control the extra muscles and to monitor the metabolic climate in the extended body bulk. Expansion of other areas, however, meant that perception and reconstruction of the real world became more and more complex. And as we know, the completeness of the world in our heads is the basis of intelligent behavior. Together with the expansion in mental machinery for analyzing and manipulating the perceived world, there must have been refinements in internal brain *organization* such as are likely to remain archeologically invisible for ever.

One of those crucial refinements in brain circuitry was the evolution of the ability to speak a complex language. Unfortunately, there is a frustrating paucity of clues from which we can infer the origins of language: a miniscule amount of direct fossil evidence, a scattering of symbolic artifacts – and a mountain of conjecture. It seems likely that a spoken language emerged slowly, from origins stretching back even longer than three million years. The explosion of new cultural patterns and the acceleration of material advance during the past fifty thousand years, which are sometimes cited as evidence of a very recent invention of language, are much more likely to stem from a more effective exploitation of what was already wired into the brain than from an improvement in the wiring itself. The biological machinery for the advance was well established fifty thousand years ago,

The increase in mean brain size from the apes through early man to Homo sapiens *is charted below. Two interesting comparisons are, above left, a part of the skull of a child* Homo habilis *fitted over the endocast of the brain cavity of an adult contemporaneous* Australopithecus boisei. *Placed in the correct position at the satures, the middle parts coincide, but the child's skull, as indicated by the extending bone at lower left, was much bigger. Although only a child it had a larger brain capacity than* Australopithecus. *Above,right, the skull of a possible* Australopithecus africanus *is compared with that of* Homo sapiens sapiens.

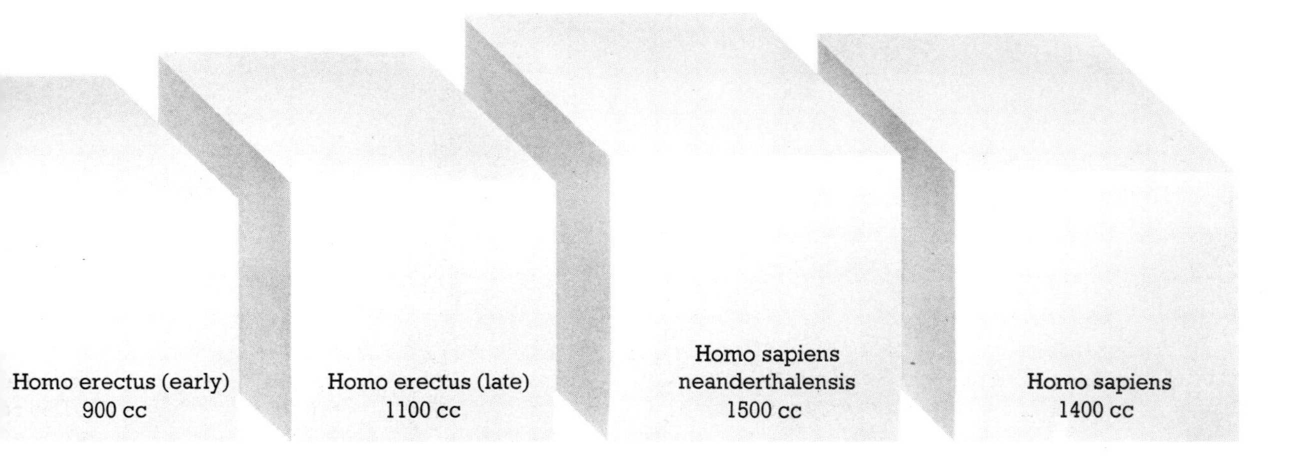

and its speed was fired by the steady accumulation of knowledge which finally hit a critical mass.

We can be absolutely sure that the ability that made language possible is firmly planted in our genes. Under anything like normal circumstances it is virtually impossible to prevent a child's learning to speak. What is more, children master what appears to be one of the most complex intellectual skills with virtually no formal tuition. From the cacophony of sounds that the infant hears it is able to construct the elements of its native language, so that by the age of five it uses perhaps two thousand words of spoken vocabulary and comprehends at least another four thousand; that these words, furthermore, are strung around at least a thousand rules of grammar adds up to a very impressive achievement.

It seems reasonable to assume that such an achievement would be unlikely if much of the structure of language were not innate. And the evidence from psycho-linguistics suggests that it is: that the structure of all languages follows the same fundamental rules; that, from the modest forty or so basic sounds (phonemes) a human being is capable of producing, the average individual has command of around one hundred thousand words. There are many thousands of different languages throughout the world. Enormously versatile and flexible though they are, the underlying unity of human languages should never be forgotten.

One intriguing question about the roots of spoken language is whether it took over when nonverbal communication (largely through gestures) became inadequate or ambiguous. Or is speech the ultimate elaboration of the vocal calls of lower primates, with gestures developing in parallel? Certainly, everyone experiences the ready recourse to mute manual gestures in a vain bid to express a point when words fail. But whether this represents an appeal to a more basic mode of communication is a highly debatable issue.

Because we can never actually *know* how our ancestors communicated, a number of biologists have sought to probe the basis of primitive communication by studying our living primate relatives, particularly chimpanzees. For reasons of history and simple practical expediency the extent of vocal communication in apes and monkeys has almost certainly been underestimated, overshadowed by the more overt and observable social interactions. Certainly, much of the communication between chimpanzees is mediated by gross bodily movements or simple touching of hands, all tending to establish and maintain social and sexual hierarchies. A hand on the shoulder of an alarmed chimp is as comforting to it as the same gesture in humans, and although soothing words are not offered by the comforting chimpanzee, soothing sounds are.

Still, apart from humans, no primate can talk. Among African vervet monkeys, an alarm system does provide one good example of unambiguous vocal communication. Three specific alarm calls – a chutter; a chirp; and a r-raup sound – mean respectively, 'watch out, there's a snake /terrestrial carnivore /bird of prey somewhere around.' Superficially, this might be seen as a first step towards using sounds to name things: does *r-raup* mean 'dangerous bird' to a vervet? This is a nice point to argue, but one is fairly safe in classifying this behavior pattern as a basic, preprogrammed emotional reaction – whereas an element of spoken language is elevating sound beyond the immediate emotional context.

It is possible to communicate through gestures alone, as sign languages for the deaf demonstrate. And it turns out that chimpanzees can also achieve a respectable mastery of sign languages. Two chimpanzees, Washoe and Sarah, managed to learn and string together in sentences more than one hundred and fifty words represented by different types of signs. Washoe learned American Sign Language; Sarah communicated by means of plastic symbols invented by her caretaker, David Premack. This compares with a frustrating six-year experiment in which Keith and Cathy Hayes tried to teach Viki, a female chimpanzee, to talk. All Viki could manage after her lengthy tuition was 'mamma', 'papa', 'up', and 'cup.'

The success with Sarah and Washoe demonstrates two important points: first, that chimpanzees can 'name' objects (that is, apply particular symbols consistently to them); and second, that they can construct meaningful (though very short) sentences. Incidentally, Washoe's ability to communicate with her tutors, Allen and Beatrice Gardner, provides a nice confirmation of the self-awareness of chimpanzees. They put her in front of a mirror and asked her (in sign language) who the chimpanzee (reflection) was. She replied, 'Me, Washoe.' These experiments emphasize the cognitive skills of analyzing and mentally manipulating objects in the environment. We can reasonably expect that our ancestors from *Ramapithecus* onwards would have been blessed with similar skills.

The chimpanzee, Lucy, being taught a language by psychologist Roger Fouts. Lucy is six years old and is able to learn about a dozen 'words' in a month. Left, she is being taught the word for 'book' and, below, the sign for 'Who are you?'

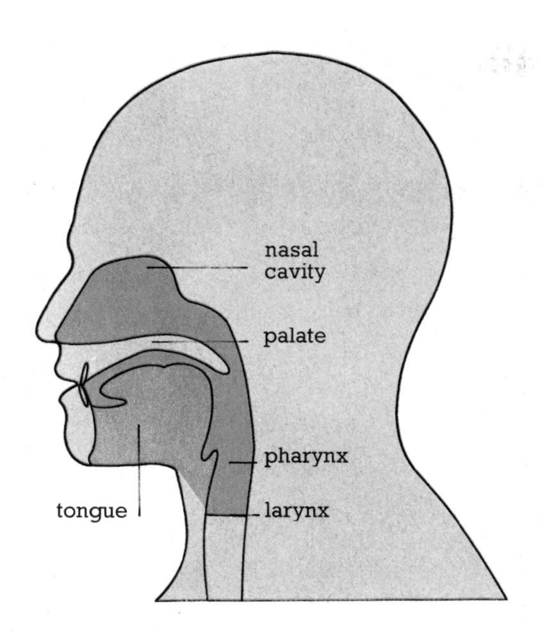

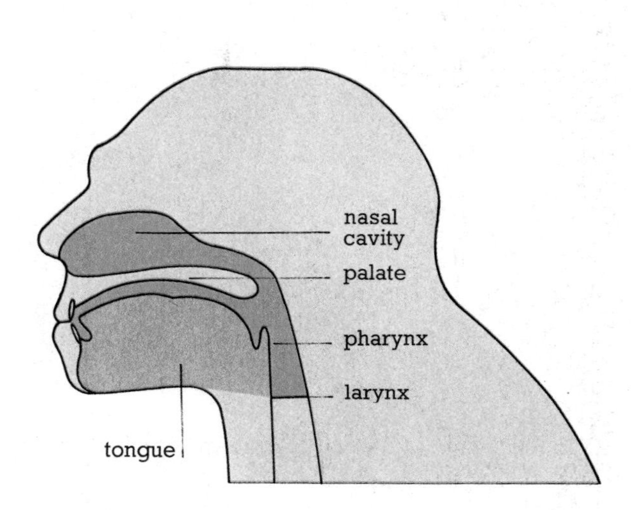

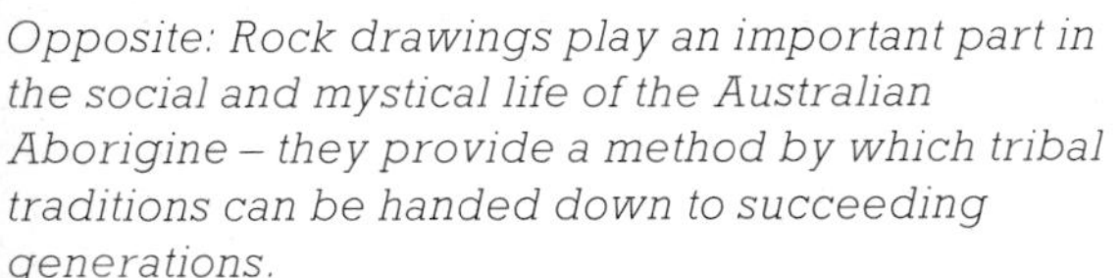
Opposite: Rock drawings play an important part in the social and mystical life of the Australian Aborigine – they provide a method by which tribal traditions can be handed down to succeeding generations.

Below left and right: Two views of the mold made by Ralph Holloway from the brain case of 1470, from which he ascertained the remarkable fact that this hominid had a larger brain than either the gracile or the robust species of Australopithecus.

Above: Man's ability to speak is, of course, a function of both the brain and the vocal apparatus. The diagram compares the size, shape and disposition of the areas used for speech in modern man with those of Homo erectus. *The reconstruction of the areas in the latter are based on the interpretation of fossil finds and comparative anatomy. It is thought that the vocal apparatus of* Homo erectus *would have enabled him to speak in a slow and rather 'clumsy' fashion.*

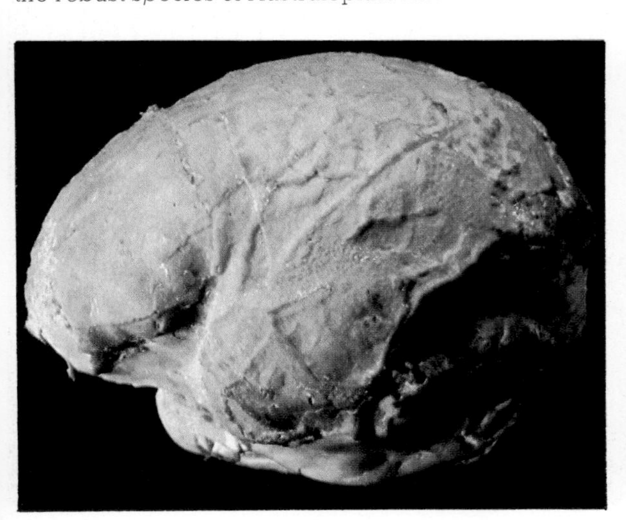

Leaving aside for a moment the question of *how* spoken language evolved, we should ask *why* it evolved. Although speech is undoubtedly a superior channel of communication, it makes good biological sense to see language as a rather useful by-product of an ever sharpening pressure to understand and manipulate the components of the environment – in other words, to think constructively. Naming objects and forming concepts are the key to this ability, thus creating a more sharply delineated world inside one's head and thus of the ability to imagine, to look back into the past, and to project into the future. The externalization of names and concepts, either in coded sounds or in specific gestures, is of course what turns thoughts into language. And biologically it is the communication of thought, rather than the thought itself, that separates humans from the rest of the animal kingdom.

Sensitive imagery is at the core of cognitive analysis of the outside world and of communication. For instance, think of a green meadow leading down to a wide, winding river edged by a thick green forest. On the river a boat is being rowed lazily by two people who suddenly laugh, causing a flock of birds to fly up, uttering cries of alarm, into the deep blue sky. Unless you are trying not to, you will depict that scene in your mind's eye; you will hear the laugh, and the call of the frightened birds, feel the warmth of the sun and smell the freshness of the grass. All this is imagery of a powerful kind, possible because words evoke pictures, sounds, textures, and smells.

Sharpening one's mental image of the world, therefore, not only enhances one's own consciousness, but also, through the medium of language, makes it possible for that consciousness to be shared. Sharing experiences through this keenly evocative channel tends to create a unique social bond. The evolution of language is so often viewed in the prosaic light of practical affairs that this basic social function is frequently overshadowed. As animals achieving new levels of socialization, the early hominids had a strong need to cope with the increased stresses that close grouping brings. Even a primitive language, which might have a 'bread and butter' function in organizing food collection, could help generate a group awareness through individuals' shared experiences. This would foster group cohesion, group identity and cooperativeness.

As ancestral communities developed an ever richer cultural fabric, embellished with social, sexual, and practical ritual, language would become increasingly crucial in weaving together threads of the cultural fabric. As contemporary 'primitive' societies demonstrate, much of the community's activity is centered on highly elaborate traditions that order the relation between individuals and families, and that nurture the younger members through the several stages heading to adulthood. It becomes incontrovertible here that language in the cultural domain is at least as important as in the world of practical affairs, and probably more so. Although we can hardly suppose that hominids of some three million years ago wove so rich a pattern of social ritual as that displayed by contemporary 'primitive' peoples, it would be surprising if there had been none at all. And by half a million years ago, with the hazy transition into modern humans, cultural tradition must have become extensive.

It is not stretching inference too far to see the beginnings of ritual in a remarkable display occasionally witnessed by Jane Goodall among her chimpanzees in Tanzania. At the onset of a violent rain storm a group of males would run repeatedly up and down a slope, brandishing branches ripped from nearby trees and calling again and again. This 'rain dance' is a remarkable prototype of symbolic reaction to the 'power' of the elements, such as forms the basis of myth and ritual in 'primitive' (and less 'primitive') peoples.

In suggesting that language has been long in the making, we place its development roughly parallel to that of tool technology, the earliest examples of which found so far come from the Lake Turkana basin in East Africa. Between tools and language there is certainly some degree of evolutionary or biological relationship: they are rooted in similar cognitive processes; imagery and structured planning are required for both. Some people have seen a link between the muscular skills used in the manufacture of stone tools and communication in the form of sophisticated gestures. When the 'vocabulary' of hand gestures had become exploited to the point of ambiguity, the argument runs, our ancestors would have been 'forced' to make the more difficult biological transition to a new channel of communication, namely that of speech. To tie the origin of language to the onset of tool manufacture, however, seems unreasonable when we consider the obvious advantages of some form of language in the basic cognitive and social developments that may have predated tool industries. It is equally valid to suggest that tool manufacture became possible because the previous evolution of a primitive spoken language laid the appropriate cognitive foundations.

But all of this is speculation. Of real evidence there is, unfortunately, very little. The Australian Aborigines have a rich social tradition, in which symbols are made richly visual in materials such as wood, feathers, colored powders, and blood – none of which would leave any archeological trace over the long periods we are considering. And the songs, dances, myths, sand drawings, incisions and circumcisions that also play a role in the Aborigines' traditions are even more 'invisible' than the impermanent tangible objects.

It is therefore not surprising that clearly identifiable symbols which might betray language are very rare in the archeological record. The cave paintings in Tanzania and in France are of course a splendid example of 'fossilized language,' but one that is relatively sophisticated dating less than twenty thousand years. What seems to be the earliest non-utilitarian object thus far discovered comes from Pech de l'Azé in France, and has been dated to around three hundred thousand years ago. The artifact, found by François Bordes, is the rib of an ox carved in a series of connected, festooned double arches – a pattern that, incidentally was still being carved as a ritual symbol some two hundred and fifty thousand years later.

One other place in which to look for signs of language, of course, is the human fossil remains themselves. And there are some interesting findings – one concerning the vocal apparatus, the other the hominid brain.

Humans are able to generate a rich variety of sounds because of the width of the throat (larynx) and the siting of the relatively short tongue at the back of the throat. Apes have a narrow throat and a long tongue, both of which tend to restrict the number of noises they can produce. We find, however, that hominids of two and three million years ago have moved significantly toward the human model, in an almost inevitable consequence of upright walking. Thus even if these hominids could not speak any kind of symbolic language, they certainly had a richer vocal equipment than their ape cousins.

Probably the most intriguing discovery comes directly from the brain itself, or rather from the imprint it leaves on the inside of the skull. In studying a number of endocasts from hominid skulls, Ralph Holloway made a specially detailed search around the bumps and muffled wrinkles on the left side. His search, of course, was for Broca's area, the brain center that organizes words into a grammatical format and initiates the muscle control required in making the precise speech sounds. During a visit to Nairobi, Holloway looked for this critical sign in the 1470 skull. Though it is more than two million years old, he found the Broca's area, immediately apparent as a small bump on the front left side of the cast.

Whether this means that 1470 could talk, it is impossible to say. But this finding does show that the channel of vocal communication was already there at a time in our evolutionary history that predates the emergence of stone tool cultures (so far as we can judge at present). Whether hand and arm movements developed merely as expressive gestures or were used to transmit specific information in everyday communication, we can still only speculate. At least one example of the robust australopithecine (a specimen from South Africa) possesses a Broca's area too, one that, although impressive as compared with the same area in a gorilla's brain, is relatively smaller than 1470's. We may infer from these discoveries that increased dependence on vocalization emerged steadily following the move of *Ramapithecus* into more open terrain, and that emerging *Homo* had a greater need than the australopithecines for a rudimentary language. All the descendants of *Ramapithecus* would therefore have been endowed with *some* vocal attributes, and these would have been more rapidly enhanced in hominids who developed further their taste for meat-eating, particularly in organized hunting and gathering. It is a fair guess that the hunters and gatherers were significantly more verbal than their contemporaries the australopithecines: they had more need to talk, and more to talk about.

We can say therefore that spoken language evolved under ecological conditions which selected for a number of complex and communal behavior patterns: such as hunting and gathering, an intricate social network, and the beginnings of tool technology. Our ability to speak is just one aspect of the evolutionary drive to create a more accurate world in our heads. Just as making and using tools allows us some measure of control of our world, so too does the ability for creative thinking and the transmission of those thoughts. This is the heart of the cognitive advancement that created the human mind.

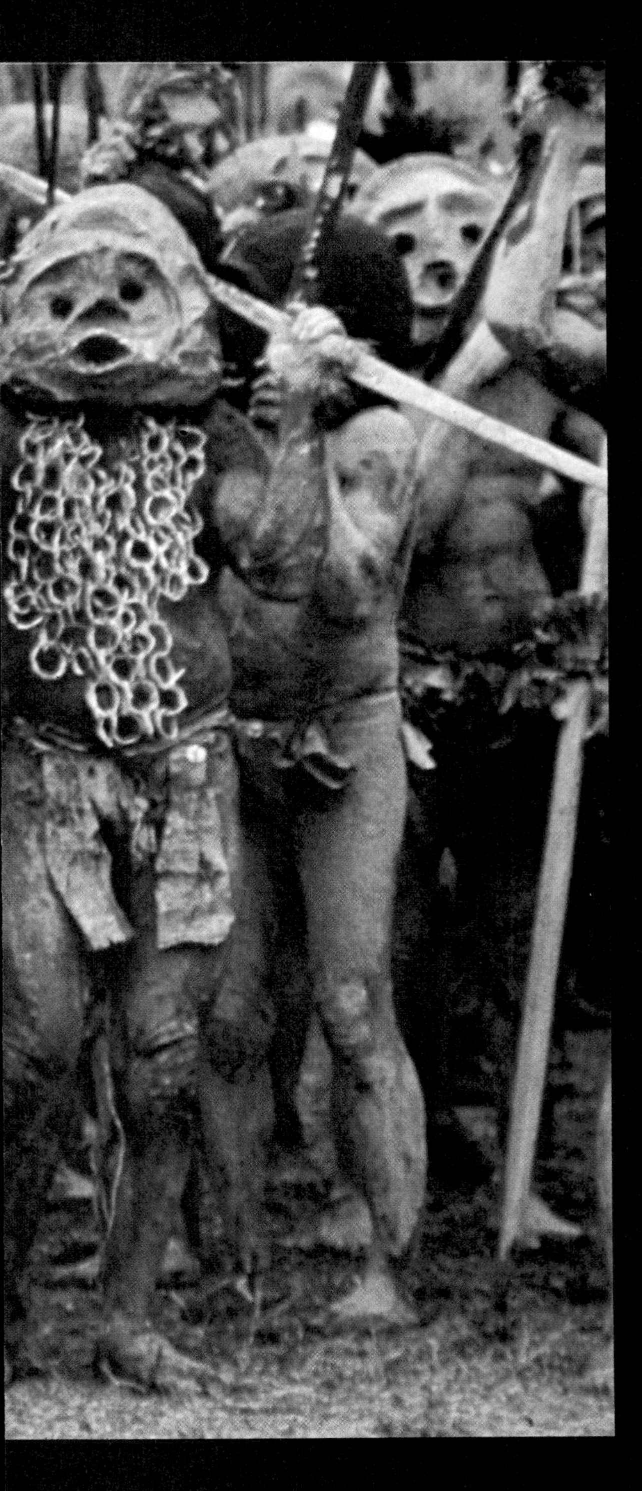

9 Aggression, Sex, and Human Nature

'The blood-bespattered, slaughter-gutted archives of human history from the earliest Egyptian and Sumerian records to the most recent atrocities of the Second World War accord with early universal cannibalism, with animal and human sacrificial practices, or their substitutes in formalized religions, and with the world-wide scalping, head-hunting, body-mutilating and necrophiliac practises of mankind in proclaiming this common bloodlust differentiator, this precarious habit, this mark of Cain that separates man dietetically from his anthropoid relatives and allies him rather with the deadliest of carnivore!' The message of these stirring words, written by Raymond Dart, is clear: humans are unswervingly brutal, possessed of an innate drive to kill each other.

On the same subject, the Nobel Prize winner Konrad Lorenz, one of the founders of modern ethology, wrote with even more eloquence: 'There is evidence that the first inventors of pebble tools – the African australopithecines – promptly used their weapons not only to kill game, but fellow members of their species as well. Peking Man, the Prometheus who learned to preserve fire, used it to roast his brother: beside the first traces of the regular use of fire lie the mutilated and roasted bones of *Sinanthropus pekinensis* himself.'

Lorenz sounded these dramatic phrases a little more than ten years ago in his celebrated book *On aggression*, the main burden of which is that the human species carries with it an inescapable legacy of territoriality and aggression, instincts which must be ventilated lest they spill over in ugly fashion. All these – the archeological evidence of cannibalism, the notions of territorial and aggressive instincts, and of an evolutionary career as killer apes – have been woven together to form one of the most dangerously persuasive myths of our time: that mankind is incorrigibly belligerent; that war and violence are in our genes.

The essentially pessimistic view of human nature was assimilated with unseemly haste into a popular conventional wisdom, an assimilation that was further enhanced by the elegant and catching prose of Desmond Morris with *(The Naked Ape)* and Robert Ardrey (with *African Genesis, The Territorial Imperative, Social Contract*, and more recently *The Hunting Hypothesis*). We emphatically reject this conventional wisdom for three reasons: first, on the very general premise that no theory of human nature can be so firmly proved as its proponents imply; second, that much of the evidence used to erect this aggression theory is simply not relevant to human behavior; and last, the clues that do impinge on the basic elements of human nature argue much more persuasively that we are a cooperative rather than an aggressive animal. If we seem to be faint-hearted by countering the proposition that 'humans are *definitely* aggressively disposed to one another' with 'no, we are *probably* a cooperative animal', then that is as it may be. There is simply no point in pretending that one has the absolute truth in the palm of one's hands, when really one is cradling a carefully-considered hypothesis.

Previous page: Asaro 'Mudmen' from the Eastern Highlands of New Guinea performing a tribal dance said to commemorate a great battle in the tribe's history when, having been driven into a river by a marauding tribe, they emerged covered in mud and scared off their attackers, who thought they were evil spirits.

For many reasons aggression looms large whenever one asks the question: what determines human behavior? But in this chapter we want to stray farther afield as well, to probe, for instance, the origins of the powerful taboo against incest, and to question the biological status of sex roles. Ever since Darwin tied knots between human beings and the rest of the animal world, many people have frantically attempted to untie them again, declaring that even though our roots are in the animal world we have left them so far behind as to make any comparisons utterly meaningless. To some extent this is true, because the quality that makes us unique in the biological kingdom, is the enormous capacity to learn.

Humans can learn virtually anything, as the rich variety of cultures throughout the world testifies. This being so, the search for the specific, stereotyped piece of behavior that is uniform in all societies would be fruitless. If there are any basic threads of human behavior – and we believe there are – they will inevitably have been richly embroidered by local cultural patterns. This is irrefutable, and it is what makes us human.

One of the most efficient pieces of biological machinery is instinct, an innately-programmed response to a specific stimulus: the herring gull chick survives because it pecks at the red spot on its parent's beak, provoking the adult to regurgitate food; male sticklebacks in the red scales that are their mating finery repel the approach of other males, or even of a red pencil that is masquerading as a rival male. Such fixed

Two examples of innately-programmed responses in animals. Above, male sticklebacks in dispute at the boundary between their territories. Left, a gull chick has pecked at the red spot on its parent's beak and thereby provoked the latter to regurgitate food.

responses exist because they are biologically appropriate and physiologically economical. The notion of instincts as powerful forces guiding animals' behavior patterns has been overestimated; and the flexibility of responses depending on prevailing environmental forces has until recently been largely ignored.

These caveats on the supposed importance of innately-programmed behavior inevitably become insistent the further we travel along the evolutionary path. One might have thought that, if nothing else,

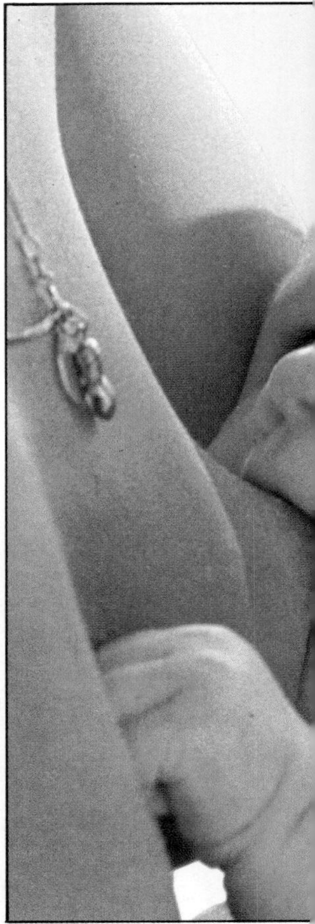

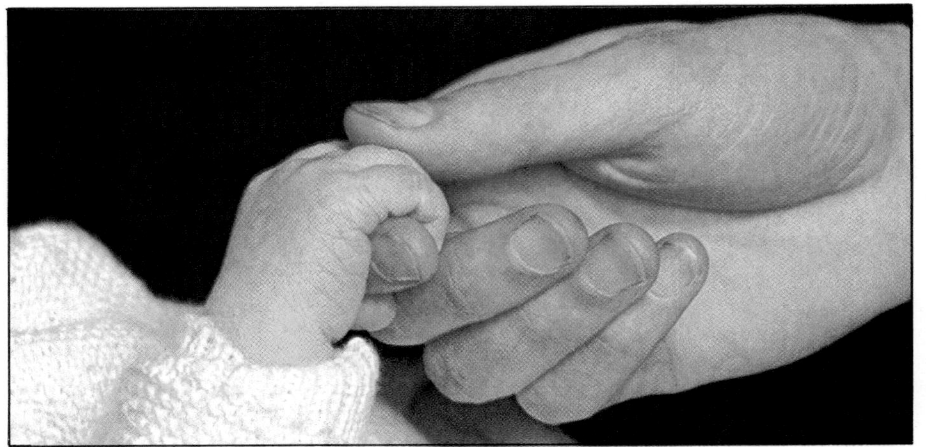

Top left: a young olive baboon clings with hands and feet to its mother's back. Left, the grasping reflex of a human baby.

Above and right: the functionally useful instinct of sucking; notice, in each illustration, the grasping fingers.

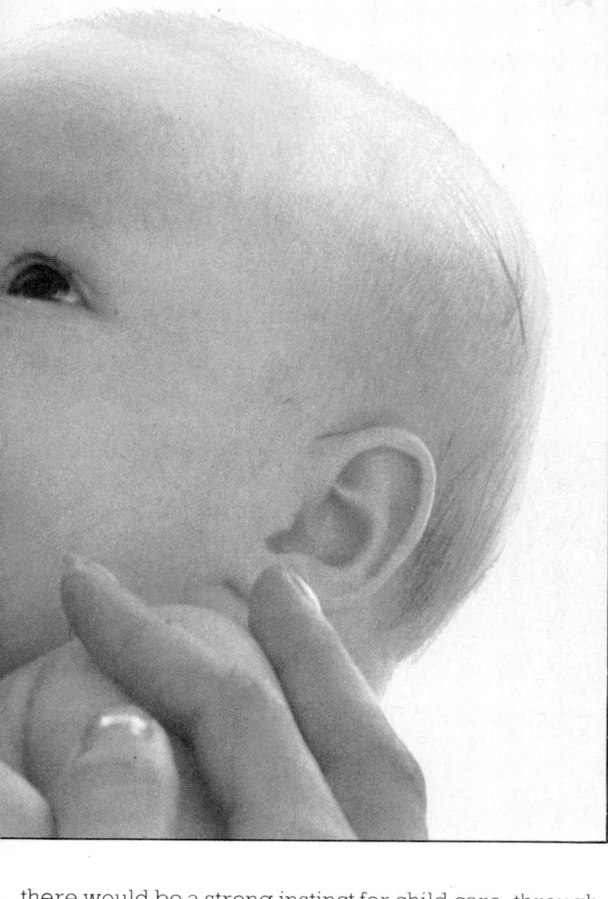

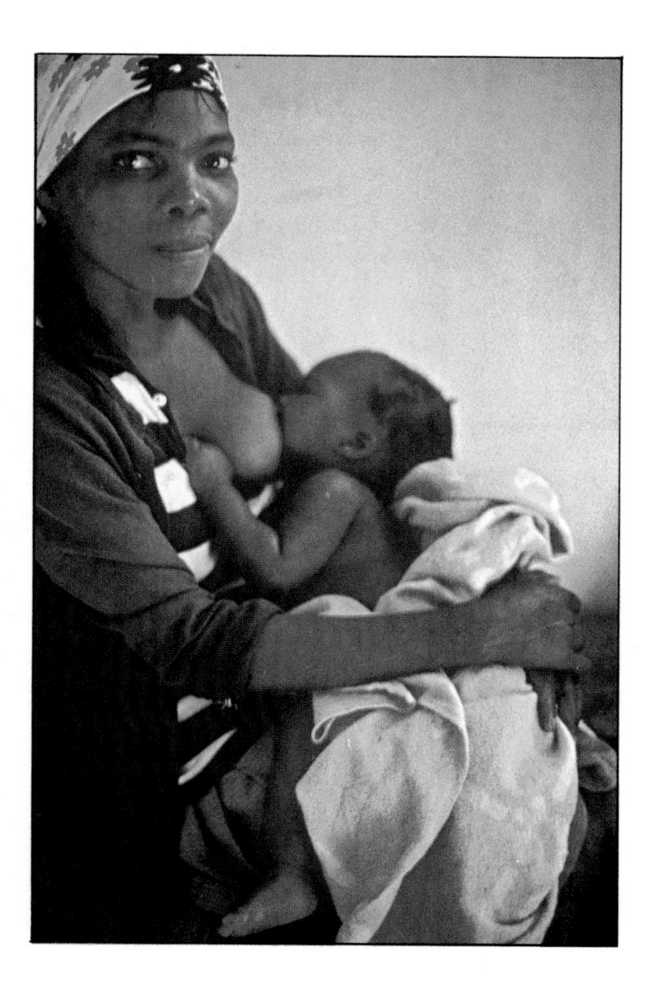

there would be a strong instinct for child care, throughout. But as the American psychologist Harry Harlow has shown, a female rhesus monkey reared in isolation instead of part of a family group becomes a very poor mother indeed; she simply does not know what to do. And Jane Goodall has noticed that chimpanzees in natural surroundings who are socially deprived through being orphaned score very low in being equipped for motherhood, at least when it comes to caring for their first offspring. On the other hand, a newborn chimpanzee does not have to learn to cling to its mother's hair: the infant is born with a grasping reflex in its hands and feet. And the grasping reflex in newborn human babies is a clear though incidental reminder of a time in our evolutionary history when we had long hair and walked on four legs, not two.

Human babies, it is fair to say, arrive in the world virtually devoid of functionally useful instincts, apart from the so-called rooting reflex and the sucking response: the baby turns its head towards the nipple and sucks. Even here, the baby's experience within its mother's womb can affect the way it later sucks at the nipple, showing the response to be more than a basic, immutable mechanism. Within a day or two after birth the baby can recognize the smell of its mother's milk. And within a week or so it has matched the sound of a familiar voice with a familiar face, showing that the long career of human learning starts very early. The rules for human behavior are therefore very simple: as each of us is a highly-sophisticated and intelligent piece of biological machinery, our responses will be finely tuned to our environment; experience and learning help us to tune these responses, but extreme experience may over-emphasize particular forms of behavior – a child reared in a home which stresses physical punishment is more than normally likely to grow up physically aggressive.

The rules for human behavior are simple, we believe, precisely because they offer such a wide scope for expression. By contrast, the proponents of innate aggression try to tie us down to narrow, well-defined paths of behavior: humans are aggressive, they propose, because there is a universal territorial instinct in biology; territories are established and maintained by displays of aggression; our ancestors acquired weapons, turning ritual displays into bloody combat, a development that was exacerbated through a lust for killing. And according to the Lorenzian school, aggression is such a crucial part of the territorial animal's survival kit that it is backed up by a steady rise in pressure for its expression. Aggression may be released by an appropriate cue, such as a threat by another animal, but in the protracted absence of such cues the pressure eventually reaches a critical point at which the behavior bursts out spontaneously. The difference between a piece of behavior that is elicited by a particular type of stimulus, and one that will be expressed whether or not cues occur is enormous, and that difference is central to understanding aggression in the human context.

There is no doubt that aggression and territoriality are part of modern life: vandalism is a distressingly familiar mark of the urban scene; we lock the doors of our houses and apartments against strangers who might wander in; and there is war, an apparent display of territoriality and aggression on a grand scale. Are these unsavory aspects of modern living simply part of an inescapable legacy of our animal origins? Or are they phenomena with entirely different causes? These are the questions that must be answered since they are so clearly relevant to the future of our species.

To begin with, it is worth taking a broad view of territoriality and aggression in the animal world. Why are some animals territorial? Simply to protect resources, such as food, a nest, or a similar reproductive area. Many birds defend one piece of real estate in which a male may attract and court a female, and then move off to another one, also to be defended, in which they build a nest and rear young. The 'choking' by male kittiwakes, the lunging by sticklebacks, and the early-morning chorus by gibbons are all displays announcing ownership of territory. Intruders who persist in violating another's territory are soon met with such displays, the intention of which is quite clear. The clarity of the defender's response, and also of the intruder's prowess, is the secret of nature's success with these so-called aggressive encounters.

Such confrontations are strictly ritualized, so that on all but the rarest occasions the biologically fitter of the two wins without the infliction of physical damage on either one. This 'aggression' is in fact an exercise in competitive display rather than physical violence. The individuals engage in stereotyped lunges, thrusts, and postures which may or may not be similar to their responses when a real threat to their lives arises, as from a predator for instance. In either event, the outcome is a resolution of a territorial dispute with minimal injury to either party. The biological advantage of these mock battles is clear: a species that insists on settling disputes violently reduces its overall fitness to thrive in a world that offers enough environmental challenges anyway.

The biological common sense implicit in this simple behavioral device is reiterated again and again throughout the animal kingdom, and even as far down as some ants: disputes over territorial ownership, and over sexual rights too, are channeled into stereotyped, non-violent competitions. This law is so deeply embedded in the nature of survival and success in the game of evolution that for a species to transgress, there must be extremely unusual circumstances. We cannot deny that with the invention of tools, first made of wood and later of stone, an impulse to employ them occasionally as weapons might have caused serious injury, there being no stereotyped behavior patterns to deflect their risk. And it is possible that our increasingly intelligent ancestors may have understood the implications of power over others through the delivery of one swift blow with a sharpened pebble tool. But is it likely?

The answer must be no. An animal that develops a proclivity for killing its fellows thrusts itself into an evolutionarily disadvantageous position. Because our ancestors probably lived in small bands, in which individuals were closely related to one another, and had as neighbors similar bands which also contained blood relations, in most acts of murder the victim would more than likely have been kin to the murderer. As the evolutionary success is the production of as many descendants as possible, an innate drive for killing individuals of one's own species would soon have wiped that species out. Humans, as we know, did not blunder up an evolutionary blind alley, a fate that innate, unrestrained aggressiveness would undoubtedly have produced.

To argue, as we do, that humans are innately nonaggressive toward one another is not to imply that we are of necessity innately good-natured toward our fellows. In the lower echelons of the animal kingdom the management of conflict is largely through ritualized mock battles. But farther along the evolutionary path, carrying out the appropriate avoidance behavior comes to depend more and more on learning, and in social animals the channel of learning is social education. The capacity for that behavior is rooted in the animal's genes, but its elaboration depends also on learning. And, as we have stressed, humans are learning animals par excellence, so we must expect that techniques for coping with potential conflict are largely learned.

For instance, among the Polynesian Ifaluk of the western Pacific, real violence is so thoroughly condemned that 'ritual' management of conflict is taught in childhood. The children play boisterously, as any normal children do; however a child who feels that he or she is being treated unfairly will set off in pursuit of the offender – but at a pace that will not permit catching up. As other children stand around, showing looks of disapproval, the chase may end with the plaintiff throwing pieces of coconut at the accused – once again with sufficient care so as to miss the target! This is ritual conflict, but it is not stereotyped human behavior. It is culturally based, not genetic. And the result is a very peaceable society.

An example of ritual conflict among adults comes from the Kurelu people in the heart of New Guinea. Superficially these people would seem to be engaged in its undoubted lethal potential they fire arrows at one another from a distance just outside the range of the weapons. Occasionally people are injured but far less often than would be inevitable if they were intent on a serious confrontation.

The ritualized nature of animals' conflict over territory is therefore the first major point to be made concerning aggression. That humans can likewise engage in ritual conflict emphasizes the biological good sense of the procedure, but does not imply that it is rooted in our genes. Humans are not innately disposed powerfully either to aggression or to peace. It is culture that largely weaves the patterns in human societies.

A second crucial feature of animal conflict is its variability. It occurs, that is to say, both between animals of different species and between individuals of the same species, and under differing environmental conditions. Anyone who argues for inbuilt aggression in *Homo sapiens* must see aggression as a universal instinct in the animal kingdom. It is no such thing. Much of the research on territoriality and aggression concerns birds. Because they usually must build nests, in which they will then spend a good deal of time incubating eggs, and still longer rearing their young, it is a biological necessity for them to protect their territory. If they did not their offspring would perish, and that ultimately implies the extinction of the species. It is therefore not surprising that most birds possess a strong territorial drive. But simply because greylag geese and mockingbirds, for instance, enthusiastically defend their territory, we should not infer that all animals do so. And it is not surprising that hummingbirds show considerably more territorial aggression than lions, even though the king of beasts is a lethal hunter. Our closest animal relatives, the chimpanzees and gorillas, are notably nonterritorial. Both these species are relatively mobile, even when there are young within the group, and they can forage for food over a wide area. Gibbons, though acrobatic, are not especially inclined to travel far; hence their need to stake out a territory, since it will contain their food supply.

The animal kingdom therefore offers a broad spectrum of territoriality, whose basic determining factor is the mode of reproduction and style of daily life. Indeed, an animal may find it necessary to assert ownership of land in one situation and not in another. Thus, the ayu, a salmonoid fish, is territorial in shallow water, whereas in deep pools it moves in close and harmonious company with its fellows. And vervet monkeys in the crowded Lolui Island of Uganda, are aggressively territorial, but at Chobe, just a few miles away, animals of this species live much more equable lives; the reason is that there they have more space, with no population problem.

That territoriality is flexible should not be surprising. It is, after all, a biological adaption to environmental conditions so that the species may survive through sufficient access to food supplies and by unhampered reproduction. If food resources and space

Above: sage-grouse cocks disputing territory. This is just one example of the ritualized displays of ownership that occur throughout some of the animal kingdom.

Right: urban vandalism takes many forms. That shown here is spray-paint graffiti on the New York City Subway.

MAIN STREET
FLUSHING
TIMES SQ
LOCAL

are scarce, then almost certainly there will be conspicuous territorial behavior. It is likewise inevitable that some individuals will fail to secure sufficient food or a place in which to rear a brood. These individuals are, of course, the weakest, and this is what survival of the fittest through natural selection really means. The pressure for selection applies with force only when resources are limited – in other words, when there is a good biological reason.

Territorial behavior is therefore triggered when it is required and remains dormant when it is not. The Lorenzians, however, take a different view: aggression, they say, builds up inexorably, to be released either by appropriate cues or spontaneously in the absence of any cues at all. A safety valve suggested by Lorenzians for human societies is competitive sport. But such a suggestion neglects the high correlation between highly competitive encounters and associated vandalism and physical violence – as players, referees, and crowds know to their cost through Europe and the Americas. More significantly, research now shows a close match between warlike behavior in countries and a devotion to sport. Far from defusing aggression, highly organized, emotionally-charged sporting events generate even more aggression and reflect the degree to which humans' deep propensity to group identify and cohesion can be manipulated.

We can say therefore that territoriality and aggression are not universal instincts as such. Rather they are pieces of behavior that are tuned to particular life styles and to changes in the availability of important resources in the environment. This being said, it follows rationally that at times of environmental stress our ancestors must have found themselves forced into territorial behavior of some kind, presumably involving competition between groups for whatever resources were limited at the time. Lorenz writes that part of such stress for early humans would have come from 'the counter-pressures of hostile neighboring hordes' – an image more dramatically evocative than it is based in fact.

There is no reason to suspect that neighboring bands ('hordes,' it may be noted, is a deliberately emotive term) were inevitably hostile. There is reason

All too common are scenes like this – a football match in Spain. It is now believed that sporting events are not a safety valve but create even more aggression.

to believe, in fact, that they would have been just the opposite, with a network of acknowledged kinship easing the contact between the separate groups.

Our ancestors of some two to five million years ago (the period when the practice of hunting and gathering were becoming firmly rooted in the fertile soil of prehuman society) would of course not have operated sophisticated kinship networks such as exist in 'primitive' people today – which may, for instance, enforce a taboo against taking a spouse with a relative's name, or a rule that marriage must be only between cross-cousins. (See Chapter 7 p. 162). But we do know that chimpanzees know who are their brothers and sisters and who are not. And we know too that chimpanzees and baboons do migrate between their various troops. The biological benefits of reducing tension and conflict between groups through exogamy appear to be almost within reach of the social structures of these our biological and ecological cousins; these benefits almost certainly would have been achieved early in hominid evolution. The notion of hostile neighboring hordes is an image born of the mistaken belief in a belligerence written ineradicably into the human genetic blueprint.

Shortage of food, either on the hoof or rooted in the ground, must nevertheless have been a cause of potential conflict between bands living in close proximity. Indeed, severe famine may well have forced hominids into belligerent confrontation with one another in open competition for the scarce food. And the band that lost out may even have ended as the victors' supper. But there is neither evidence nor any reason to suggest that hominid flesh, either roasted or raw, appeared on our ancestors' diet specifically as a source of food in any but the most extreme circumstances. A much more likely consequence of conflict over food resources, so far as can be judged from what we know of both animals and present-day hunter-gatherers, would have been the dispersal of bands and even the temporary scattering of the members of the individual bands, a practice that ensures the best use of the limited food that is available.

To most 'civilized' people the thought of cannibalism is repulsive, so that when individuals find themselves confronted with no alternative but to eat the flesh of their fellows or starve – as has happened a

Overleaf: Chimbu men engaged in ritual battle in the village of Mindima, New Guinea.

number of times in recent years following a plane crash in a remote place – the rest of the world looks on with horror, as well as more than a little morbid fascination. Conventionally, cannibalism is seen as one of the most barbaric practices of 'primitive' tribes, and as a motive behind their 'savage aggressiveness'. So, when archeologists find in the fossil remains of our ancestors, such clear signs of cannibalism as charred and cracked limb bones and skulls with a hollowed-out base, the idea that we evolved from an innately-depraved and savage stock is readily swallowed. In this context Raymond Dart's suggestion that our ancestry was rife with predation and cannibalism appears to put a seal on that gory view of the animal within us. But to be seduced by this argument is to ignore the real significance of cannibalism.

Apart from lions, humans are the only mammals who on occasion deliberately eat each other. When a male lion wins control of a pride, he will often consume the young cubs and set about producing offspring of his own, working hard to prevent other males from inseminating 'his' females. Ruthless and wasteful though it may appear, the biological reason for the dominant male's behavior is evident: the offspring produced by the pride will have been sired by a very powerful animal, providing a brutal but efficient method of natural selection. Cannibalism in humans, however, takes place for different reasons.

Broadly, there are two sorts of cannibalism and the distinction between them is crucial. First, there is the eating by members of one tribe of individuals in another – usually as the end result, or even the motive, of an aggressive raid; such is the conventional version of the practice, and it is known as exocannibalism. In the second form, known as endocannibalism, people eat members of their own tribe. The motives for the two forms of cannibalism are unmistakably different, and it is equally significant that their occurrence is tied to the distinction between hunting-gathering and agricultural peoples.

Both forms of cannibalism have been practiced in all parts of the world where 'primitive' people thrived untouched by contemporary norms of behavior. No tribe, whether it indulges in endo- or exocannibalism, includes human flesh as a regular part of its diet. Eating that flesh is not a way of getting a meal; primarily it takes place as part of some kind of ritual. And almost never does that ritual embody naked, unbridled aggression. Even among the infamous tribes in the highlands of New Guinea, who appear to launch cannibalistic raids for the fun of it, the context is one of extensive tribal ritual. Months of preparation – weaving symbolic adornments and the carving of elaborate wooden images – precede a raid, and it is abundantly clear that the entire exercise has a powerful unifying effect on the tribe. It may be deplored that a socially-binding ritual should be geared to killing members of another tribe, but for the participants those deaths are in a way almost incidental to the group's total involvement in a common goal. The habits of the New Guinea tribes are, in any event, extremely rare, and as against cannibalism manifested in this extreme form we may set the other extreme, in which people swallow a small morsel of a dead relative as a mark of love and respect.

Exocannibalism is much more often an accompaniment of hostilities than the cause of them. For instance, the Theddora and the Ngarigo of southeastern Australia formerly cut off and ate the muscles from the arms and legs, and the skin and flesh from the sides of the body of their victims – all the while uttering declarations of contempt and scorn for the dead man. And the Sumo, a tribe of South American Indians, chopped up and ate the bodies of enemies killed in battle, an activity designed to insult the dead victims as well as, it would seem, to prevent the possibility of their being able to inflict harm on the murderers.

By contrast, endocannibalism at its simplest may be the customary way of laying the dead to rest. The Kallatians, an Indian tribe, ate their deceased relatives, and looked upon the practice of burning bodies as barbaric. Among the Dieri, an Australian aborigine tribe, the fat was cut away from the arms, legs, face and belly of someone who had died – the dissection being usually done by an old male relative of the dead person – and then eaten. These people believed that the fat possessed important powers which might be acquired in this way, but they also wished to assimilate the personality and soul of the dead person.

This wish for continuity is the most common motive of endocannibalism. Evidently, though, there have been many different notions about which parts of the body contained the transmittable essence of the dead person. A number of South American tribes, for instance the Amahuaca, Jumano and the Pakidai, believed the bones to be the seat of the soul. When someone died, the remains were burned and the ashes were then mixed with drinks, ensuring that the life that had passed from the dead person was now located in those living beings who drank the mixture. The Chiribichi roasted their dead, the purpose being to collect the melting fat as it dripped from the body; this was then drunk, thus preserving the dead person's life in the bodies of his friends and relatives.

Signs of cannibalism in the archeological record may therefore indicate an act of hostility (as in exocannibalism) or be a fossilized token of love and respect. How can we know which? Though the answer is inevitably that we can never be sure, there is one interesting clue in a comparison of cannibalism in South American tribes. From a survey of fifty-four tribes in this area it turned out that sixteen indulged in endocannibalism and the rest were exocannibals. Of those sixteen, fourteen were hunter-gatherers. And of the thirty-eight exocannibals, only six were non-agricultural. In other words, of all the hunter-gatherers in the survey seventy per cent practiced endocannibalism and only thirty per cent exocannibalism. Hostilities, and the associated exocannibalism, are much more closely associated with the sedentary life of agriculture than with the nomadic little bands of hunter-gatherers.

Farmers have much more reason than nomads to instigate battles: their social groups usually constitute a larger agglomeration of people; they have land and crops to protect; and what is more, when the harvest has been gathered they have longer stretches of relatively free time in which to wage war. Briefly scanning the archives of human warfare, we discover that warriors more usually waited until the harvest was in before launching battles, free time and the need for a store of food being obviously important in the strategy of war. By contrast, although hunter-gatherers' day-to-day lives are not arduous, they must engage continuously in a mixture of work and leisure, and there are no long intervals when they might find themselves with nothing to do.

It may be added that, lacking time in which to elaborate the rituals of warfare possible to agriculturalists, the rituals of hunting people surrounding social maturation, initiations, and so forth, are similarly far more restricted.

The thirty per cent of South American hunter-gatherers in the survey who did occasionally practice exocannibalism may well have learned the habit from their agricultural neighbors. And the fact that the vast majority of the hunters are not exocannibals in an area where this form of cannibalism is not uncommon demonstrates cogently how incompatible the habit, and the linked hostilities, are with the hunting way of life. It is not unreasonable to suppose that in the absence of agriculturalists exocannibalism may be very rare indeed.

It is in this light that we should view the cracked and charred bones and the chipped skulls from the Choukoutien caves near Peking (Chapter 6, p. 132). Almost half a million years ago a band of our ancestors held the heads of dead people in their hands and carefully enlarged the hole through which the spinal cord entered the brain. They then scooped out the brain, not an easy task through this awkward entrance and almost certainly ate it. Was it out of malice or respect?

The caves will hold forever the secret of what this brief, but fortunately preserved, episode of human history really means. But we believe that the Choukoutien skulls speak to us of the concern and respect our ancestors had for each other, as among those hunter-gatherers who practice endocannibalism now. By contrast, chimpanzees finishing off a meal of young baboon will unceremoniously crush the skull to reach the brain tissue they so obviously relish. The Choukoutien ritual feast could possibly have been the culmination of a bloody encounter between two bands of *Homo erectus*, the brains being consumed to gain power over the enemy and the skulls preserved to be scorned and derided. But if we are attentive to the evidence from all available sources this interpretation is much less likely than the alternative of endocannibalism, enacted as a sign of group awareness, identification, and the desire for continuity.

Incidentally, whether as endo- or exocannibalism, the relics of such ritual behavior in our fossilized history are a clear indication of self-consciousness, the awareness of life and death in terms of self and of others.

Altogether, then, the notion that humans are inherently aggressive is simply not tenable. We cannot deny that twentieth-century humans display a good deal of aggression, but we cannot point to our evolutionary past either to explain its origins, or to excuse it. For that is what the equating of territorial aggressiveness in the animal domain with waging war in the human one often amounts to – an excuse. The fallacy of thus adducing our animal origins should now be evident. Wars are planned and organized by lead-

ers intent on increasing their power. And they are fought usually by people not driven by an innate aggression against an enemy they often do not see. In war men are more like sheep than wolves: they may be led to manufacture munitions at home, to release bombs, or to fire long-range guns and rockets – all as part of one great cooperative effort. It is not insignificant that those soldiers who engage in fierce and bloody hand-to-hand fighting are subjected to an intense process of desensitization before they can do it.

War is a battle for power over people and for resources such as land and minerals, neither of which are relevant in hunting and gathering societies. With the growth of agriculture and of materially-based societies, warfare has increased steadily in both ferocity and duration, culminating in our current capability to destroy even the planet: powerful leaders have found more and more to fight about, and increasingly effective ways of achieving their ends. We should not look to our genes for the seeds of war; those seeds were planted when, ten thousand years ago, our ancestors for the first time planted crops and began to be farmers. The transition from the nomadic hunting way of life to the sedentary one of farmers and industrialists made war possible and potentially profitable.

Possible, but not inevitable. For what has transformed the possible into reality is the same factor that has made human beings special in the biological kingdom: culture. Because of our seemingly limitless inventiveness and our vast capacity for learning, there is an endless potential for difference among human cultures, as indeed may be witnessed throughout the world. An essential element of culture, however, consists in those central values that make up an ideology. It is social and political ideologies, and the tolerance or lack of it between them, that bring human nations to bloody conflict. Those who argue that war is in our genes not only are wrong, but they also commit the crime of diverting attention from the real cause of war.

This last criticism applies even more strongly to those who cite innate aggression to explain violence within nations, particularly in the overcrowded urban areas. There are many reasons why a youth may 'spontaneously' smash a window or attack an old lady, but an inborn drive inherited from our animal origins is certainly not one of them. Human behavior is extraordinarily sensitive to the nature of the environment, and so it should not be particularly surprising that a person reared in unpleasant surroundings, perhaps subjected to material insecurity and emotional deprivation, should later behave in a way that people blessed with a more fortunate life might regard as unpleasant. Urban problems will not be solved by pointing to supposed defects in our genes while ignoring real defects in social justice.

The Porgaigas of New Guinea still practise cannibalism. They do not live in villages but in isolated family huts scattered over many square miles. A few months before this photograph was taken two men were eaten.

One supreme biological irony underlies the entire issue of organized war in modern societies – the cooperative nature of human beings. Throughout our recent evolutionary history, particularly since the rise of a hunting way of life, there must have been extreme selective pressures in favor of our ability to cooperate as a group: organized food-gathering and hunts are successful only if each member of the band knows his task and joins in with the activity of his fellows; a good deal of restraint on natural impulses during the stalk and capture of the prey is likewise essential. The degree of selective pressure toward cooperation, group awareness and identification was so strong, and the period over which it operated was so extended (at least three million years, and probably even longer), that it can hardly have failed to have become embedded to some degree in our genetic makeup.

We are not suggesting that the human animal is a cooperative, group-oriented automaton. That would negate what is the prime evolutionary heritage of humans: their ability to acquire culture through education and learning. We are essentially cultural animals with the capacity to formulate many kinds of social structures; but a deep-seated biological urge towards cooperation, towards working as a group, provides a basic framework for those structures. The cooperative behavior born of our hunting-gathering heritage combines with the long-established social nature of primates to galvanize social units with an extraordinary ability to tackle environmental challenges. This is, of course, why we were so successful as an evolving species. But as we now know, the evolutionary endowment of group cooperation, nurtured in small socially-cohesive bands on the savannas of Africa, can be exploited by powerful leaders to motivate people by the hundreds of thousands to go to their death on the battlefields of the world.

Unfortunately, it is our deeply-rooted urge for group cooperation that makes large-scale wars not only possible, but unique in their destructiveness. Animals that are essentially self-centered and untutored in coordinated activity could neither hunt large prey nor make war. Equally, however, massive warfare would not be possible without the inventive intelligence that has produced the increasingly sophisticated hardware of human conflict. It is therefore as unhelpful to blame the scourge of war on our cooperativeness as it would be to blame it on our intelligence. To do either is simply to evade the real issue – those ideological values on which nations are based, through which governments manipulate their people.

One aspect of ideology that has had its share of conflict and manipulation is sex – specifically the phenomenon of incest taboos, and the question of sex roles. The former has for many years had sociologists, anthropologists, and geneticists scratching their heads in puzzlement: is the taboo a social tool or an expression of a deeply-ingrained biological law? And the question of sex roles can hardly be raised without provoking abusive attacks, and reopening what has degenerated into an irrational debate. The issue here is also one of biological as opposed to social origins.

It is a simple rule of thumb that behavior which is genetically determined will be universal among human populations, and will appear unerringly in the presence of an appropriate environmental cue. By contrast, social activities that are cultural inventions will not be universal and they are usually governed by cultural rules. Unfortunately, the avoidance of incest among humans fits neither category – or rather, somewhat perplexingly, it fits both: the prohibition of incest is virtually universal, and there are social rules to insure that it is observed (transgressors are usually punished, sometimes with alarming ferocity).

Commenting on this embarrassing paradox, the anthropologist Claude Lévi-Strauss has said, 'We are faced with a series of facts which are not far removed from a scandal. . . . Here therefore is a phenomenon which has the distinctive characteristics both of nature and its theoretical contradiction, culture. . . . It therefore presents a formidable mystery to sociological thought.' Lévi-Strauss, taking firm hold of the dilemma, resolves the paradox by declaring that 'the prohibition of incest is where nature transcends itself.' If by this he means that the incest taboo was once rooted in a simple biological function, later branched into a social role, and now blossoms in the context of social tradition, he is probably correct.

The one certain feature of the incest taboo is that arguments over both its functions and the mechanism of its expression are tortuous in the extreme. The least hazardous way of disentangling them is to consider the various issues separately, beginning with the reasonable assumption that within the period of relatively recent human history at least, the avoidance of incest was advantageous. Although sexual activities and marital alliances are not synonymous, when one talks about the social rules that govern incest one is really most concerned with rules that limit marriage. The fact that sexual relations between kin may be transiently tolerated, whereas marriage between the partners would not, merely emphasizes the social role of the taboo in maintaining the structure of the society. It does not imply that views on sexual intercourse between kin and the requirements for marriage outside the immediate family are unrelated issues.

What, then, are the possible benefits of avoiding long-term sexual relations with one's immediate kin? First, there is a biological case to be made. Put simply, the argument here is that close inbreeding inflicts genetic dangers on the offspring. Harmful recessive genes that would remain unexpressed while dispersed thinly in an out-breeding population would become concentrated in an inbreeding group, and the defects would then become physically manifested. Certainly, there is evidence from inbreeding experiments on animals that these fears of incest are valid: reduced fertility, shortened life expectancy, increased susceptibility to disease, and puny stature all occur in inbred animals. And from the few systematic studies of inbred human populations, we can be sure that the same kind of defects occur there as well.

There are, however, at least two reasons for believing that the incest taboo did *not* arise from a biological need to avoid inbreeding. The first is that the results of inbreeding one sees either in animal experiments or in human populations have occurred against a short-term time scale. Undoubtedly, if a mixed population does harbor potentially dangerous recessive genes then inbreeding *will* bring them to the surface, producing defective individuals. But if a population continued inbreeding over a very long period of time, it is more than likely that the dangerous genes would be sifted out: the affected individuals would prove less fit than their fellows and so would perish.

An inbreeding population is in fact a very efficient system for enhancing beneficial genetic traits. Indeed, some geneticists argue that our ancestors could not have evolved as rapidly as they apparently did *unless* they engaged in incest. A beneficial alteration in the organization of chromosomes, they suggest, would be slow to establish itself in an outbreeding population; it might even be diluted out, and thereby lost forever. Such advantageous changes in inbreeders would, by contrast, be rapidly fixed as a trait giving them a possible advantage in the continuing battle for survival and thus fostering evolutionary progress.

As a possible explanation of the incest taboo, the dangers of inbreeding do not appear particularly persuasive. The opposite view – that there are benefits in outbreeding – is however, much more attractive. Although it is not quite universal, reproduction by sexual methods, from the most insignificant plant all the way through to sex-obsessed humans, is in biological terms by far the most successful one. Although there are evident disadvantages in demanding that two individuals be in the same state of sexual readiness at the same time and in the same place, these are greatly outweighed by the benefits of genetic variability that are offered by sexual reproduction. Even plant breeders know that in order to produce a new variety they have to cross two different plants; it is simply not possible to achieve the same effect by manipulating the chromosomes of a single parent plant. The genetic variability provided by mixing chromosomes is the basis of evolutionary success as much as it is for the commercial success of plant breeders. In the natural world a richer scheme of genetic variability is achieved by mixing chromosomes from unrelated individuals than through mating with close kin.

The benefits of outbreeding, rather than the hazards of inbreeding, must therefore be a strong candidate, biologically, for the reason behind the human incest taboo. We are not suggesting that the rich variety of incest taboos now to be found in different parts of the world should be viewed as social ritualizations of this biological function. It seems possible, however, such a function is indeed at the root of the prohibition against human inbreeding, and equally that this single thread underwent extensive elaboration in the biological and social evolution of our ancestors.

If, as many people believed until recently, avoidance of incest was absent or rare in the animal world, our entire argument would collapse. But it is not. Among social animals, where the opportunity for incest exists potentially, there turn out to be a variety of social systems whose effect is to minimize that opportunity. In red deer, for instance, the chances of incest are drastically reduced because for most of the year the sexes remain segregated. Only during the mating season do the males associate with females, and they are then extremely intolerant of the other males with whom they normally associate. Since the young join the single-sex groups after they mature, the opportunities for incestuous matings are meager.

In mixed sex groups the problem is trickier. A common pattern of group living is the harem, in which a single male governs a small number of females who are accompanied by their young. Less dominant and juvenile males wander around in bachelor bands, usually waiting their chance to become a harem leader. In such social groups, as they are formed, for instance by the zebra, by patas monkeys, and hamadryas baboons, the dominant male keeps a watch over 'his' females, thus effectively preventing all forms of incest apart from that between father and daughter, of course. He is unlikely to get a chance at this, however, because one of the young nearby bachelors usually sneaks in and abducts the newly-matured female. In zebras the abduction occurs when the female shows her first visible signs of sexual readiness. But bachelor hamadryas baboons kidnap their victims long before any sexual bond is possible. The young male has to act as a 'foster parent' until his bride is sexually mature.

Social systems such as these thus prevent incest by the simple expedient of separating the potential incestuous partners. The social grouping in which kin mating is most likely is the one comprised of adult males and females, plus their maturing offspring, over a long term. Good examples of this are rhesus monkeys and chimpanzees, both of whom carry on promiscuous lives not notably harassed by jealousies or firmly-asserted rankings. Offspring know which female is their mother, but because of the promiscuous sex relations they can have no idea which male sired them. Father-daughter incest is therefore almost certain to occur, and it is difficult to see how, biologically, a specific taboo against this could arise.*

* The fact that of all forms of human incest, that between father and daughter is by far the most common, may not be unrelated to this aspect of our primate past. Incest between a mother and son is the rarest form with sibling incest somewhere in between.

Be that as it may, rhesus monkeys do avoid mother-son and sibling incest because the young male round the time of sexual maturity, often moves off to another troop. But field researchers have seen that even those males that remain in their native troop rarely mate with their mothers; and sibling pairing is likewise much lower than that between unrelated animals. A similar pattern occurs in chimpanzees. Jane Goodall was able to observe the first flush of sexual maturity in a young female, Fifi. When she came on heat the young chimpanzee was greatly interested in her new powers, and eagerly solicited sexual advances from all the males around her – except from her brothers. They did try to mate with her, and succeeded, but only to the accompaniment of loud protestations. After this the brothers, Faben and Figan, rarely tried again.

Why do the brothers and sisters, and the mothers and sons, not mate when they have the opportunity? Some aspect of the dominance relationship may explain why sons rarely try to mount their mothers. Generally males are sexually successful only with females lower in the overall order of dominance. As a mother retains a kind of dominance over her son, this may explain his inhibition against mating. But what of siblings? Why should there be any barrier to copulation here? Perhaps their long association in growing up as

part of the same family erodes any interest in the sex of their potentially incestuous partner, simply through familiarity. Primates are essentially inquisitive creatures, constantly seeking new stimulations, so that although childhood companionship forges strong familial bonds it also blunts the keenness of the sexual drive.

The question of the mechanism of incest avoidance in these animals and in our early precultural ancestors is of course vital to understanding the entire issue. Interestingly enough, a small number of contemporary societies practice infant marriage, giving us an opportunity to assess the notion that 'familiarity prevents breeding.' The intended spouses are brought together when still children, and are then reared as brother and sister until they are ready to be married. Commonly such marriages are described as embarrassing or boring, and sexual activity is typically pursued with less than the usual enthusiasm.

We are therefore faced with the possibility of an innate prohibition against incest, which operates through a dampening of sexual inclinations towards

One method of avoiding incest is the harem– shown here is a dominant male Hamadryas baboon watching over 'his' females.

Elaborate wedding ritual, seen here in the Korokoro people of north-east Kenya, is often an important element of social custom and usually embodies mechanisms for avoiding incest.

an individual with whom one has shared the experience of childhood. If one contemplates a biological origin of incest avoidance in animals at least, this kind of mechanism is much more acceptable than one demanding that blood relations should instinctively recognize the fact – a point that Freud was much exercised about.

If we now move toward the arena of human society, we can begin to see how the basic biological tendency to avoid incest may have become modified and have accrued new functions. The virtual universality of incest taboos is worth stressing again, since it emphasizes the possibility that an aspect of social culture could have been generated out of more deep-seated biological functions. And certainly the tradition is very deep-seated in the human race. Asked why they practice incest taboos, 'primitive' peoples are more likely to say that its violation would be to invite the wrath of the gods, or some such personal or tribal disaster, than that it enhances genetic variability or that outbreeding helps to forge new alliances between tribes. The point is that the taboo has become firmly ritualized. On the other hand, Margaret Mead encountered an unusual degree of objectivity in a member of the Mountain Arapesh in New Guinea who said to her, 'If you married your sister you would have no brother-in-law. With whom would you work? With whom would you hunt? Who would help you?'

Such comments almost certainly point to some of the social roles of the incest taboo that have served to plant it even more firmly in the repertory of human behavior than it is in that of animals. Whatever way it was perceived, the prohibition eventually became formalized as central to the rules of marriage. One may wish to ponder on the meaning of the existence of rules governing an activity that is supposed to be 'natural'. Some people argue, indeed, that the very existence of rules against incest implies that there is a natural drive to commit the deed rather than avoid it! However, against the background of evidence from animal societies and the very real social advantages that would have begun operating at least several million years ago, the argument appears somewhat perverse. The prohibition is formalized in ritual because we are a cultural animal.

In contemporary hunter-gatherer societies the operations of the marriage rules which avoid incest become very obvious: as we have noted (Chapter 7 p.162) one of the longest single journeys a !Kung person makes is to find himself a bride. The !Kung bands are widespread, as is typical of hunting societies, and so the long trek becomes unavoidable if a spouse is to be found from outside the home band – marriage rules insist that he must.

Now let us turn our minds back to our ancestors of some three million years ago. We find them beginning to organize themselves into hunting bands of around twenty-five individuals – the number so common in contemporary hunters. Very probably before meat eating had become more than just an occasional habit, these ancestral bands would have been larger, more like the fifty or so in troops of modern baboons. And these groups would not have been as widely dispersed as they would become when hunting emerged as an important social and economic activity. The migration of young males from one group to another would therefore have presented no special problem. And even those who remained in their home group would have a fair chance of finding a partner who was not either a sibling or very closely related.

As a hunting and gathering way of life slowly established itself in prehuman societies, incest avoidance through migrations would become more of a problem simply because the distance to be traveled was greater. At the same time, however, the need for migration might have escalated, again for a simple reason. In a group of twenty-five individuals there might be perhaps fourteen offspring, about one-quarter of whom would be reaching sexual maturity at any one time. The problem, then, is clear: lack of choice. And any imbalance in the sex ratio of the maturing children would exacerbate the difficulty. Migration might be the only answer, and the individuals with the greater inclination to migrate would have a better chance of surviving and producing offspring of their own. Forming sexual alliances outside one's immediate group – exogamy – might, therefore, have become biologically necessary, as well as biologically advantageous because of the basic issue of genetic variability; and both developments would serve to ingrain the behavior pattern in the species' genes.

As we know from contemporary societies, there are important consequences of exogamy in the social and economic spheres. Kinship links make for peaceable contacts between neighboring bands and are often instrumental in forming temporary alliances for large-scale cooperative ventures, such as hunts. With their increasing mental powers our ancestors must at some time have become aware of such advantages – an enlightenment that might have spun the first threads of a formalized taboo, which eventually became interwoven with a powerful inclination into a social ritual. The need for the taboo as such would have been sharpened if sexual maturity arrived before a youth was ready to find a spouse (whether by transferring to another band in a hunter-gatherer society or, after agriculture had been invented, to a neighboring village).

There are many possible social benefits of pursuing incest avoidance besides those already mentioned. For instance, by seeking a sexual partner outside one's own group, competition between related males can be avoided. And a father who produced offspring through his daughter would find himself with two women in his household. This might at first appear to be a satisfactory arrangement in hunter-gatherer societies, since the females provide most of the food. But in practice, and probably for reasons connected with social status as much as mundane practical economics, a male usually attracts more than one spouse if he is an especially good provider of meat, a product with at least as much social cachet as nutritional value. This, of course, is as much an argument against polygamy as it is against father-daughter incest, but we should not be surprised to find intertwining threads in the complex fabric that makes up human cultural life.

We therefore see the incest taboo as having fundamental biological benefits, and as having accrued social advantages so crucial that it was enhanced through formalized cultural patterns. The taboo is certainly a deeply-rooted, long-established social tradition. It is therefore interesting to speculate on what might happen if, tomorrow, an all-powerful committee were to wipe out all existing traditions and replace them with new ones, but were accidentally to omit mention of the incest taboo, or of something like it. Very probably, in industrial communities bearing no relation to the small hunter-gatherer and agrarian societies, the omission would go unnoticed. An observant sociologist might record that sexual relations were more common outside the family than within, and this would be a tangible manifestation of the behavioral mechanisms (mediated through familiarity) that evolved to minimize incest in social animals. In short, without the need for social and economic alliances typical of primitive societies, there would be no requirement for a rule of exogamy.

We turn now to another aspect of human sexuality

that is likewise an extraordinary mixture of biological and social factors – the nearly universal social and political dominance of men over women. This dominance may manifest itself either in the strictly formalized cultural organization of many eastern countries, or in the widely-accepted social norms of western suburbia. The origins of the differential status of the sexes are, of course, currently in hot dispute, with battlelines drawn up at opposite extremes: from the nativists who claim a purely biological base for male 'superiority', with characteristic sex hormones and psychological propensities enforcing the differences, to the environmentalists who insist that women are exploited and oppressed by a male conspiracy, and that social conditioning is the mechanism for perpetuating an unjust status quo.

If an aspect of behavior is universal or nearly so in human societies, we should at least allow ourselves to suspect some kind of genetic basis for that behavior. On the other hand, there is a great deal of flexibility in the relation between men and women in differing societies – emphasizing the importance of other factors in orchestrating those relations. What we want to do here is look at the male/female issue from an evolutionary perspective in an effort to determine whether any general principles concerning sex roles are to be found. Before beginning to do so, however, we should stress that even should it appear that true differences between men and women had been inserted into our genes through the process of selection this cannot be used to bolster a continuation of social and economic inequalities that are embedded in so many cultural traditions. What was biologically sound a hundred thousand or two million years ago is not to be necessarily equated with social justice today.

One of the most obvious physical differences between men and women, apart from specifically sexual characteristics, is size: men are taller and more muscular. They also have narrower hips than women. All of this means that often males are stronger and can run faster than their mates. At its simplest this distinction amounts to a neat analogue with our close biological and ecological cousins, chimpanzees and baboons. And ecological and environmental pressures are particularly crucial in determining sex-linked behavior.

Taking our cue from contemporary hunter-gatherer societies – and there seems to be no good reason why we should not – we have already suggested that the actual hunting of game was done exclusively by males,

Other than sexual differences there are physical variations between men and women: as a general rule men are stronger than women and can run faster. The couple shown here are pygmies from the equatorial forest of Zaire.

while the females probably gathered important plant foods. Considering this division of labor, it is probably not without significance that the intellectual functions governing so-called visual-spatial abilities develop much earlier and are more deeply ingrained in boys than in girls, whereas girls' verbal development tends to outpace that of boys. It may be, of course, that these somewhat controversial observations are the result of the different upbringing of boys and girls.

The consistency of these psychological differences with the separate social roles of males and females in hunter-gatherer societies is, however, intriguing; hunters do depend on an adequate visuo-spatial sense; and gatherers, who would also be child-minding, would inevitably be much more likely to be talking, both to each other and to the children, an important aspect of which is their education. If these different psychological facilities can indeed be verified, they may well be a tangible heritage from our evolutionary past. It is unlikely, at any rate, that the disparity is dramatic enough to interfere significantly with the aims of a society dedicated to sexual equality.

Division of labor, as we have repeatedly stressed, is the key to human social organization. But it does not *inevitably* follow that simply because tasks have to be divided between members of a social group they will become established as sex roles. Indeed, as we know from various contemporary 'simple' communities,

The meat upon which Eskimos largely subsist is hunted by the men who therefore play a dominant role in their society. The Eskimo shown here is eating Muktak *– the outside skin and blubber of a whale.*

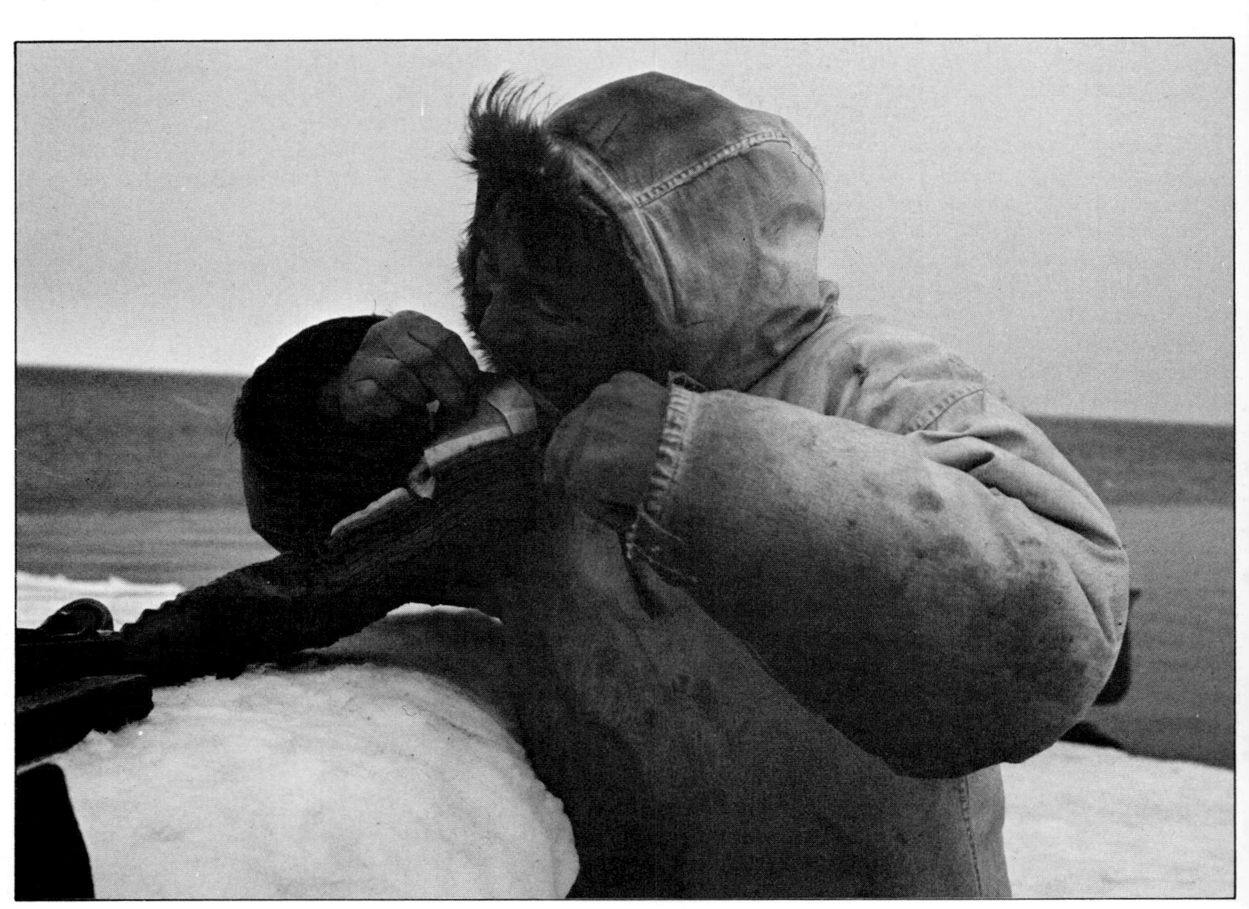

functions regarded as 'male' in one society may be thought of as 'female' in another. On the other hand, the universality of hunting as a male preserve is striking. It is this single factor that appears to dominate the social status of the different sexes, specifically through the way the spoils of the hunt are distributed.

Why is it that females do not engage in large-scale hunting? Even with a possible lesser degree of development in visuo-spatial skills, women could certainly have become proficient hunters; for the crucial trick in hunting is in approaching very close to the prey, rather than indulging in a long and arduous chase. To the question of why no women in contemporary hunter-gatherer communities become hunters, the answer, inescapably, is to be found in child-bearing – a biological role that women are, of course, stuck with. And so it is for the period of early infancy during which the child suckles from the breast. In the !Kung of the Dobe area, a mother may suckle her child up to the age of four years. And a woman who is either pregnant or still suckling a child would hardly be able to engage in long hunting trips. The prohibition is compounded still further by the inescapable obligation for her to go on bearing and rearing children throughout her reproductive life. In the !Kung, as we have noted, the four-year period of close childhood dependency corresponds to the frequency with which the women have babies. That a woman in this society must expect to go on having children is due to two reasons: the high rate of infant mortality and the small number of women in the band. All of this means that the population will decline unless they all produce and rear their quota of offspring.

The importance of hunting in establishing social structure is not in the activity as such, though a particularly skillful hunter will attract much admiration. Rather, it is in the value placed on the meat that is its product. Once again, the importance of meat is not so much that it is especially palatable or nutritious, but rather that it is vital to cultural interchange between individuals within bands, and between bands too. In a hunting and gathering community the exchange and sharing of food is at the core of the social structure, but there is an essential difference between the fate of hunted meat and gathered plants; plant foods generally remain within the immediate family circle whereas meat may be distributed outside the immediate families of the men who have provided it.

The distribution of a prestigious and valued commodity such as meat confers considerable social and 'political' status on the giver. The lines of such distribution are usually guided by kinship, and there is an expectation that gifts will be reciprocated. A person who is in a position to distribute meat is therefore at the center of a network of reciprocal relationships which help to strengthen alliances between the groups involved. And the distributors of meat are those who hunted it – the men. Women in hunter-gatherer societies are therefore, in this respect, at a disadvantage so far as their social position is concerned.

Meat, then, serves as a valued currency in the political affairs of developing human society. As we have mentioned, there is a wide variation in the amount of meat consumed by hunting peoples, determined largely by the ecological setting: in tropical regions plant foods form a major part of the diet, whereas on the arctic tundra the Eskimos eat virtually nothing but meat and fish. And in the political milieu of hunting societies it may be observed that the greater the importance of meat, the greater the dominance of men over women.

Roughly, four levels of hunting and gathering may be distinguished. First, there are people such as the Hadza of Tanzania and the Paliyans of southwest India, who eat little meat and derive most of their food from plants; these they collect and eat individually rather than share. Even before they were moved from their homelands because of encroaching civilization, the Hadza men were not particularly enthusiastic hunters, and so therefore there was little in the way of political currency to maintain the structure of social interchange.

A second level is one at which, although hunting is an important activity and one that carries much social prestige, it supplies less than fifty per cent of the diet. The !Kung, for instance, though they are very definitely hunting people, eat perhaps thirty to forty per cent of their diet in the form of meat. Between the two levels of society already mentioned is another, characterized by some degree of cooperation between the men and women in their hunts, which usually involve chasing the prey into nets, without the long treks engaged in by the !Kung. The BaMbuti live this kind of life to some degree, though the men occasionally kill elephants in a most courageous, solitary manner.

The fourth type of community is typified by the Eskimos. They eat only meat, and it is caught exclusively by the men, on whom the women – whose tasks

Above: a family of pygmies in the Central African Republic.

Examples of division of labor in mankind: left, Turkana fisherman by the lake and, left, a woman preparing ugali *– a corn dish.*

are concerned with processing meat and skins – are completely dependent.

The status of women in these hunting societies is inversely proportional to the amount of meat hunted and to their involvement in the hunt. For instance, among the Hadza and the BaMbuti there is something approaching social equality between men and women, with little concern about overall status. Men and women are equally free to choose their spouses, to take lovers, and to separate if they so desire. Among hunters of the !Kung type, in which power gained through distribution of meat sets apart the men from their women, there is a much greater awareness of status. The man with the greatest status is the one who has shown himself to be a skilled and resourceful hunter – and who is therefore much sought after as a potential groom. Among traditional communities like those of the Eskimos, women suffer badly on the social scale, being treated largely as sex objects with little control over their own fate.

The picture that emerges is clear in outline, though in detail there will be variations from one local culture to another. Remembering that from about one million years ago, our ancestors began their migrations from Africa into the colder regions of Europe, we can guess that their diet placed less and less emphasis on plants and correspondingly more on meat. The impressive slaughter sites at Terra Amata and Torralba, where many large animals met their death, testify to the competence of *Homo erectus* at hazardous big-game hunting as much as half a million years ago. If, as seems likely, these hunts were primarily carried out by men, then we can suppose that the social milieu would have been *at least* as male-oriented as the !Kung's, and probably much more so.

It was almost certainly through facing the exigencies of big-game hunting and of coping with less than hospitable climates that these ancestors of ours made the final transition to *Homo sapiens*. The question we have to ask is, how much of the probably long-practiced exercise of male social dominance thereby became rooted in our genetic make-up? The variability in sexual status among contemporary hunter-gatherers reminds us of the behavioral flexibility that may go into cultural adaptation to ecological conditions. But it is perhaps worth reflecting that in none of these societies do women achieve more than a rough equality. And only rarely do women in nonhunting societies rise to the male equivalent of social superiority and political power. Agricultural peoples display a greater diversity of social organization than hunters; but as with hunters, the men frequently have control over prestige items, such as ritual plants that are used in social exchanges. Thus, they acquire the social kudos that gives them power over their women.

If our cultural expression of sex roles were totally unbridled by basic biological patterns and influences, we should expect to see a more even balance between male and female-dominated societies. It would surely be somewhat remarkable if all the social and economic systems that have ever been invented were to have led to male social dominance simply through cultural conditioning, as the extreme environmentalists suggest. The resilience of even deeply-rooted traditions should, however, not be underestimated, and it may be that those traditions have been strong enough to remain roughly constant, thereby weathering the transition from hunting to an agricultural, and from an agricultural to an industrial society.

Although meat represents only thirty to forty per cent of the !Kung's diet, they are hunting people and the prowess of the men in this field sets them apart from the women. Here a hunter is poisoning his arrows.

As with incest taboo, the cultural expression of some degree of male dominance may have become so deeply embedded in our social norms as to appear to be strongly biologically based. But, again as with the incest taboo, it seems more probable that subtle biological mechanisms do exist that incline social behavior towards male dominance, although these require strong environmental features to encourage their full expression. Certainly the housewife who sits in her beautifully appointed suburban home contemplating whatever new item her husband's large income has most recently provided, thereby adding to the family prestige, is much more akin to the Eskimos than to the Hazda, and is apparently treated as such.

Admittedly there are now many formidable and deeply entrenched barriers in the way of women who wish to participate equally in the political and economic life of our society. These barriers are constructed on a foundation of social prejudice and conditioning. But they remain so intractable because a long evolutionary history has almost certainly implanted within us a *propensity* towards sexual differentiation of status, with males dominating. Having said as much, we should stress that with childbearing no longer a major obstacle in the way of a woman attempting to pursue a long-term career, there is nothing to prevent a total restructuring of society so that men and women may contribute equally to the domestic economy. A greater participation in political affairs would soon follow, perhaps to our advantage – as would also women's freedom to take lovers and to separate from their spouses.

To say this is simply to assert that cultural determination is a more powerful social force than any of those biological propensities which still linger in our genes. If we wished to change the structure we could do so, without any fear of some primal urge welling to the surface and sucking us back into some atavistic pattern. Vested interests might be loud in their protest, but our deepest biological nature would not. We are, after all, the ultimate expression of a cultural animal; we have not totally broken free of our biological roots, but neither are we ruled by them.

10 Mankind in Prospective

Previous page: A triumph of human ingenuity – a view of the surface of Mars as received on earth from the Viking 1 lander.

On 20 July 1976, with the clocks of the eastern seaboard in the United States standing at precisely 07.53, a small, tripod-like spacecraft touched down gently on the surface of the planet Mars. When the dust had settled, with the craft's smooth metal gleaming in the sun, the scientific instruments it contained began their task of probing the mysteries of the Red Planet, for so long the object of scientific curiosity as well as myth. Indirectly, the human mind had for the first time journeyed from its native planet to take up long term residence in another part of the solar system.

The Viking Project, which eventually put two landers on Mars, was a triumph of human ingenuity, and determination. Here was arresting proof of the unrivaled power of the human intellect, the product of millions of years of biological evolution. But the very success of the project points up sharply the dangerous paradox that the human species must now face: intellectually we seem able to tackle the most challenging problem with confidence, whereas we find ourselves at the same time, in dealing with basic human issues, pathetically inept. It is perverse in the extreme that hatred, prejudice, and conflict should thrive in a world made materially sophisticated through the almost limitless inventiveness of the human mind.

For instance, as the Viking 1 lander began its truly awe-inspiring analysis of the Martian environment, the world's newspapers rippled with tremors of global conflict. The focus of the disturbance seemed innocent enough: the Olympic Games in Montreal. A threat of disruption hung over these games because of the continued and unforgivable abuse of a fundamental tenet, the equality of mankind. Specifically, the violation of that tenet lay in supposing that people with lighter skins are superior to those whose bodies are more conspicuously pigmented. And a few weeks after the games, bloody rioting exploded in the country where this supposition is most blatantly violated: South Africa. The outburst of violence and killing that took place in the South African township of Soweto in the summer of 1976 is just a mild foretaste of what the world in general and South Africa in particular can expect in the future if this vile artificial division between people is maintained.

Without doubt the distinction that is made between so-called whites and so-called blacks has generated one of the most serious threats to long-term peace in our world. Quite apart from the sterile and vacuous arguments about supposed intellectual disparities between whites and blacks, the division of humanity into these rigid categories is total nonsense in itself. There are virtually no truly black or truly white people. Certainly the degree of skin pigmentation differs between populations in different parts of the world. The function of pigmentation as a protection against the potentially harmful ultraviolet rays of the sun demands that it should be so; as one moves nearer to the equator the penetration of ultraviolet increases, heightening the need for protection. It is therefore to be expected that in general, long-established populations living near the equator will be more pigmented than those farther away. This, however, produces different shades of brown, not just black and white.

To be heavily pigmented in an environment that is exposed to a high level of ultraviolet radiation is an expression of biological harmony with that environment, and can in no sense be a comment on intellectual or social attainment. As early populations moved north into colder climates the need for heavily pigmented skins diminished, and so complexions became gradually lighter. As migrating populations moved down through North America and into South America the need for protection once again reappeared, and once again pigmentation increased. The fact that in general the skins of equatorial Americans are not as dark as those of equatorial Africans is probably a consequence of the relatively short time in which increased pigmentation has had to re-evolve on that side of the globe. Degrees of skin pigmentation among different populations therefore reflect adaptations to their different physical environments. The social mobility of the twentieth century has of course cut across these adaptations, and it has caused problems for all concerned: when light-skinned people travel in hot countries the sun readily takes its toll on their unprotected skin, in spite of even the most costly sun creams; and when heavily-pigmented people live in sunless climates they may have to add more vitamin D to their diet as it is made less efficiently in their shielded skin.

Rioters burn a beerhall and a bus in Soweto township near Johannesburg – the aftermath of the police opening fire on and killing demonstrators.

The tendency to term certain populations as black while grouping the rest into a white club is, therefore, grossly inaccurate. And it is more than a matter of pedantry to object to the terms, for the separate groupings are exploited to open up social and economic gulfs, with the 'whites' on the right side and the 'blacks' on the wrong, although there exists no basis for such a division. With the label of 'blacks' ready at hand, it is just too easy for a 'white' person to attach it to any 'appropriate' group of people, thus attributing to them a collection of supposedly global characteristics while retreating behind the security of one's own convenient label. Such an exercise is no more than an effective technique for ignoring the realities of the world and replacing them with inflexible prejudice. There are no global characteristics of either 'whites' or 'blacks,' for the reason that these groups as such do not exist. There is, however, the global characteristic of belonging to the human species with perhaps five million years of *Homo* evolution behind each of us.

The nonsensical use of the terms 'whites' and 'blacks' *must* be dropped as a first step toward breaking free from the divisive thinking behind it. The current social and economic status of the world's populations, which sees a minority of light-skinned people taking a major share of the earth's resources, is the result of historical development, much of which lacked any vestige of human dignity and justice: the political and economic imperialism of the past cannot be used to argue its continuance today. Certainly, that dominance has no long-term future. If the divisiveness continues it will cut through to the heart of humanity,

and finally destroy it. The choice is simple – and it is stark; either the true brotherhood of mankind is universally recognized, whatever the degree of skin pigmentation, or the future is very bleak indeed.

Fundamental though it is, this issue of human conflict is but one of many in a world in precarious balance on the thin edge of the threat of global war as the supposed guarantor of global peace. War, on a scale both large and small, has etched itself deep in the annals of human history, and there are those who declare that because of our biological heritage such conflict is inevitable. This unfounded claim is based on a misunderstanding of the nature of such basic animal behaviors as territoriality and aggression, and an equal ignorance of the forces that nurtured the evolution of humanity. In this respect there is a certain irony in the

Equatorial pigmentation. Top right, a lighter-skinned Xingú Indian from the Mato Grosso region of Brazil; right, Maasai herdsman of Kenya; left, two young Xingú women being led away from a hut where they were kept in seclusion for almost two years as part of a ceremony connected with the onset of menstruation. The women have lost some of their pigmentation and their skin is several shades lighter than normal.

These segregated walkways leading to the platforms of a railway station are one aspect of apartheid in South Africa.

spectacular success of the Viking mission, since its target, Mars, has so long been identified with the god of War!

If it were truly the case that the human animal is unswervingly propelled toward open conflict with members of its own species, then the prospects for long-term tenure of technological society on earth would be very poor indeed. With the means of mass destruction at hand, it would be merely a matter of time before their use dealt a blow to life on earth from which it might never recover. Those who argue that humans are innately aggressive must therefore be pointing to the inevitability of such an event. Unfortunately, it does not necessarily follow that a humanity not harboring inbuilt aggressiveness will avoid reaching the same end, as our recent history and current political posturing so clearly warns us. The possibility of a massive international confrontation is real.

How, then do we explain international conflict and the real threat of global confrontation within the context of the human community in which there is no powerful undercurrent of aggression waiting to surge to the surface?

The search for human origins in the dust-dry strata laid down by ancient lakes and streams is an attempt to answer that question. It cannot give us the whole answer. No single line of reasoning can do so, for the reason that the processes of human evolution have produced so very complex a creature. But it can, we believe, yield some very fundamental clues. The many hundreds of researchers who are now at work unearthing fossilized bones, studying the fate of modern skeletal remains, examining stone-tool technologies, and learning about the social and economic life of contemporary 'primitive' peoples, are all sketching, line by line, a steadily-growing picture of the forces that nurtured the emergence of mankind. It is this overall picture, not just the stones or bones by themselves, that will give us the insights we need.

Evolution has endowed human beings with two remarkable characteristics, each of which alone would make us very special animals: first, the enormous capacity to learn about and interpret the world around us; and second, the ability to structure and manipulate the environment in arbitrary ways so as to create culture. Add these two together, combine them with a degree of social cooperation found elsewhere only in some insect societies, and one finishes up with an extraordinary product: an animal with the potential to achieve virtually anything.

The human infant comes into the world less well equipped than virtually any other animal offspring to deal with the world it is thrust into. We carry a meager handful of instincts into the world with us, but we do have a voracious appetite to learn, an appetite that begins to be fed almost from the moment of birth. As individuals we are very much an expression of the home and immediate social environment in which we grow up. Our immediate social community is characterized by material, ideological, and social patterns that are part of the 'local' culture. What of the world population to which all local communities belong? What universal aspects of behavior can we expect to see?

Precisely because evolution produced an animal capable of tackling whatever challenge the environment might offer, the answer must be that very few behavioral patterns are rigidly built into the human brain. Obviously our brains are not jumbled networks of nerve cells with no overall structure. The *anatomy* of the human brain is very well ordered, but it is built in such a way as to maximize *behavioral* adaptability. Within reasonable biological limits, humans, it is fair to say, could adapt to living in almost limitless numbers of ways. Indeed, this flexibility is manifest in the rich pattern of cultures expressed throughout the world.

Throughout the later stages of human evolution, from about three million years onwards, there was, however, one pattern of social behavior that became extremely important, and we can therefore expect the forces of natural selection would have ensured its becoming deeply embedded in the human brain: this is cooperation.

Primates, particularly the higher primates, are all social animals: they live in groups and engage in com-

The human baby is helpless compared with the offspring of many other animals – the young kongoni, for example, is far better equipped to deal with the world around it.

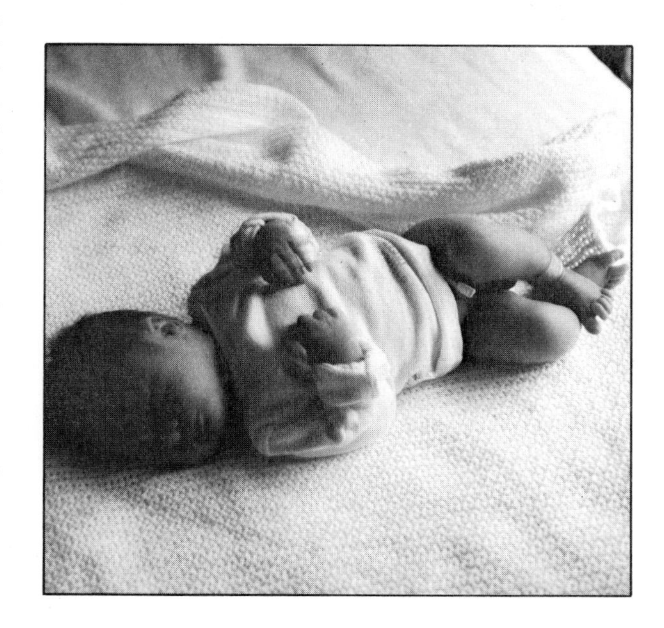

1

2

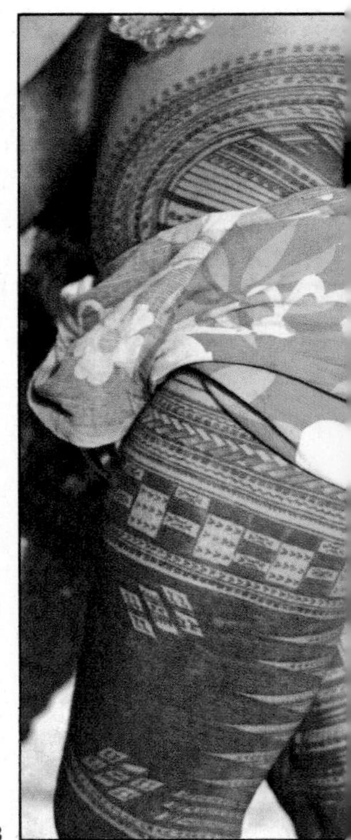

3

5

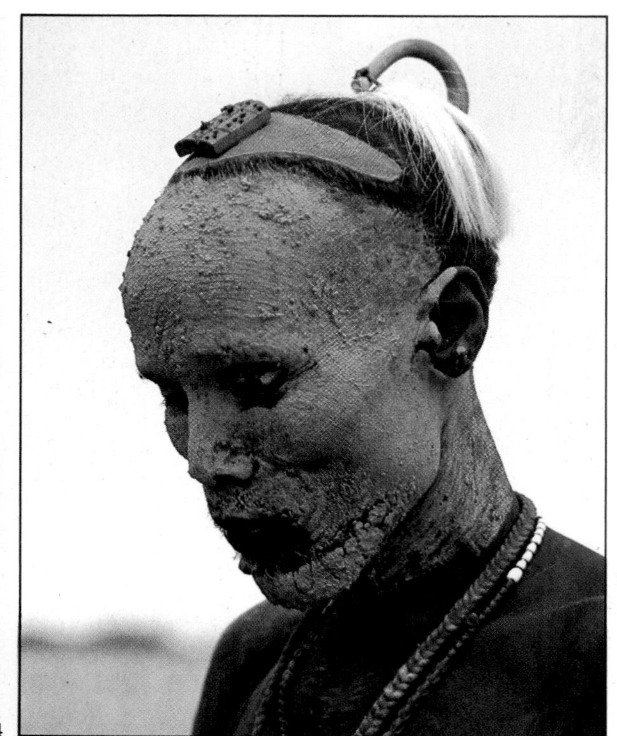

4

6

Self-adornment is a method of group identification. This particular form of ethnic expression can be in bodily decoration or in dress, and those shown here are: 1. a Mekeo warrior from the south east of New Guinea; 2. two Peulh tribesmen, Mali; 3. body tattoos on a Salamunu from Upolu Island in Western Samoa; 4. a Wapenamanda warrior from the central highlands of New Guinea; 5. a tribesman from southern Ethiopia; 6. traditional costumes in the South Tirol, Italy.

plex social interactions. It is, therefore, not surprising that, as primates, humans too are social animals. But unlike any other primates we extend social behavior into the patterns of subsistence. The core of human evolution was the social group focused on a food-sharing economy: plant foods and meat were brought to a home base and shared. Without a keenly developed sense of cooperation the social organization needed for day-to-day division of activities between members of the group, which is the basis of the mixed economy, could not have worked.

Because the mixed economy of hunting and gathering brings with it a much more efficient exploitation of resources in the environment and also sharpens the edge of social interactions – both of which enhance adaptability of the human animal – evolutionary forces favored its development. It is probably the single most important factor in the emergence of mankind. And cooperation was an essential element of its success. More than any other piece of social behavior, the motivation to cooperate in group effort is a direct legacy of the nature of human evolution.

To insist that blind cooperation is a universal aspect of human behavior would, of course, be to negate the flexibility and independence of the human mind. What is in us is a very readily-tapped tendency for group identification and group endeavor. Most communities have social rules and customs (the stuff of culture) that provide a framework through which 'groupness' is expressed. The rules and customs of different communities may vary, and ethnic groups may express their identification through such material channels as self-decoration or domestic architecture. But, throughout, the goal is the same: the sense of belonging to and therefore contributing to the group. We all experience such an urge, and it may extend from the need to be 'accepted' by a group as small as perhaps three or four people at school, in college, or within the local community, all the way to supporting a national sports team along with tens of thousands of other fans.

It is possible to argue that this urge for group identification is at the root of much of the conflict we have seen in the world: war would not be possible if people were not inclined to rally round their flag and fight for the good of the country, whatever that 'good' may be. To say, however, that the biological heritage of group identification and cooperation is the *cause* of war would be the same as claiming that guns are the cause of war. Both simply represent the *means* by which war is waged. War only became possible when there was something to fight about; and here we must look back to the agricultural revolution, the transition from hunting and gathering to farming.

Because of the nature of the hunting and gathering economy, groups were small, containing around twenty-five people. In order to survive these people must have been finely tuned to their environment, knowing what plants were ripening when and where, and having a detailed knowledge of the habits of possible prey animals. They must have remained in balance with the world that fed them, otherwise the following year they would have starved. As we know from contemporary hunters and gatherers, this form of economy is both extremely efficient and remarkably secure, providing sufficient food with a minimum of effort: the !Kung people, for instance, work just three days of the week in an area which is marginal in the extreme, and so have a great deal of time in which to weave rich patterns into their cultural fabric. As with all hunter-gatherers the !Kung have to remain essentially mobile, and therefore accumulate very few material possessions.

The economy of hunting and gathering therefore implies living in small bands, in which there is no advantage to be gained by appropriating the territory of a neighboring band; the practicalities of exploiting it will be more costly than the 'prize' is worth. Undoubtedly there must have been times in which scarce localized resource led to open conflict, but in general the hunting and gathering economy militates against such confrontations. It was not until village settlement became possible on an extensive scale that neighboring populations would have had 'good reason' to covet their neighbors' resources. In contrast to the hunting and gathering way of life, where small bands are best suited to exploiting the food resources, the concentration of food through farming allows local populations to grow. Villages can become towns. Now, if one farming village decides it will take over the crops of a neighboring village it benefits because its own population can then expand because of the extra food. First, of course, there would be the little matter of a battle, but as long as the losses in the confrontation are not too great, the first village is now at an advantage: its numbers grow at the expense of its neighbors.

With the advent of permanent settlements comes the birth of materialism as well. Sedentary life in villages allows the accumulation of non-essential objects, and it is with such objects that marks of status and wealth are often associated. Experience tells us that accumulation of wealth breeds the desire for more as often as it satiates the appetite. The phenomenon is more than simply collecting items for their own sake; it is a kind of psychomaterialism, a phenomenon akin to the lust for power. And, once again, power is possible only when there is a large body of people over which to exercise it. Clearly, the possibilities of seeking, holding, and expanding power were much greater after the agricultural revolution than before. And there are two basic routes to the expansion of power: skillful political maneuvering, or successful military operations.

It is a mere ten thousand years since agriculture first became established, and from that momentous change have flowed the industrial and technological societies in which we now live. Because that period is so short in biological terms, we can be sure that the brains of the hunter-gatherers ten thousand years ago were exactly the same as ours today: their experience was different, but the mental equipment with which they analyzed their world was identical to that inside twentieth-century human skulls. For three million years we were hunter-gatherers, and it was through the evolutionary pressures of that way of life that a brain so adaptable and so creative eventually emerged. Today we stand with the brains of hunter-gatherers in our heads, looking out on a modern world made comfortable for some by the fruits of human inventiveness, and made miserable for others by the scandal of deprivation in the midst of plenty.

A world in which two-thirds of its people starve while the rest bathe in material security, and go on seeking still more, must be lacking an essential degree of global compassion and justice. Poor countries may suffer doubly at the hands of their affluent neighbors: in commercial exploitation the terms of trade mean that primary food producers can never catch up with their customers economically, and it is to the 'benefit' of the customers to ensure that they do not. Often, food producers find it more profitable to sell their produce to Europe and the U.S., where it is used to fatten livestock, rather than channel it to the poorer population of their own country. It is salutory to remember that your juicy red steak may be partly the product of a very inefficient conversion of foodstuffs diverted from the mouths of the underfed people in the suffering two-thirds of the world. How, remembering, can you justify that?

Affluent countries may inflict suffering on the poorer communities through incompetence as well as by design. The unthinking way in which Western medicine is foisted upon societies that are totally unsuited either to operate it or cope with the fruits of its 'success' – those fruits being an expanded population – is potentially as harmful in the long term as economic exploitation. The technologically advanced must try to forget the notion that they have all the answers, and that what is good for them is good for everybody. This is not necessarily true. Indeed, it may well be that the economic and social strategy which has brought wealth to many Western countries is in the long run a certain recipe for global disaster. The high standard of material affluence enjoyed in Europe and the U.S. may well be *too* high to be feasible on a world-wide scale: the drain on the planet's resources, and the accompanying rape of the environment, could be too great to sustain a population of even four thousand million (the present level) for more than a hundred years or so. And it is inconceivable that the differentials that currently exist between the rich and poor countries can be part of a long-term strategy of economic survival. If those differentials persist, global tensions will rapidly pass exploding point.

One issue that ties together the questions of material resources and of standards of living is the global planning of energy. We can now be certain that at the current rate of usage the world's reserves of oil and gas will be exhausted well within this century. We know, too, that there is enough coal to supply our energy-hungry world for many centuries. But it may also turn out that in order to survive we will have to leave that coal buried and unburned. The problem is that, as with other fossil fuels (oil and gas), coal when it is burned releases carbon dioxide. Over the long term, a large build-up of carbon dioxide in the atmosphere could alter the world's climate sufficiently to disrupt seriously the pattern of agriculture. Moreover, agricultural production in the U.S., China and the Soviet Union is so finely balanced with demand, particularly for grain, that even the slightest *normal* variation in weather can cause havoc. The results of a global warming of 2.5°C, the anticipated outcome of a dramatic rise in the carbon dioxide level within sixty years, are virtually unimaginable.

Major policy decisions will have to be taken within the next thirty years if this prospect is to be diverted. And those decisions will be useless unless they are

1

2

3

4

5

Domestic architecture depends upon the availability of local materials, but is also an expression of 'groupness'. The examples shown here are: 1. the tents of Moroccan nomads; 2. an Eskimo completing an igloo in the North West Territories, Canada; 3. a grass hut in Mali; 4. a homestead in the Sologne area of France; 5. a house on stilts in Lau Lagoon, Solomon Islands.

agreed upon globally: there would be little point, for instance, in the Soviet Union deciding not to unearth its massive coal reserves if the U.S. goes ahead and burns its coal, or vice versa. Undoubtedly the question of future energy resources is about to strain the machinery of international decision-making as it has never been strained before. There is no mistaking that if the wrong decisions are made within the next thirty years, human life on earth could be set on the downward spiral toward extinction.

Intimately related to the energy issue is size of population. There is a simple equation that says that, with limited resources to exploit, the life span of modern human society is inversely related to the number of people in that society. In other words, the more people there are, the sooner resources are used up. With the exception of the sun's rays, there is no resource that is not limited in some important way. True, new future technologies are certain to exploit materials in ways not dreamed of at the moment. And it is possible that new technologies could guide the human species through novel material societies for many thousands of years. But the key to successful exploitation of new material resources is that the process should be unhurried, so that the maximum potential should be squeezed out of the material world. With the population due to double in the next thirty years, and with the period of doubling tending to shorten, there is no prospect of such an unhurried approach.

The pressure of enormous population expansion will undoubtedly stimulate many innovations designed to cope with the billions of mouths to be fed, innovations that might have remained unborn in the absence of the stimulus. But who can stand up and confidently say that the rate of innovation will match the rocketing demand? Currently the gap is large, and it is widening minute by minute.

One obvious solution to the problem of a large world population is to have a small one instead. It is quite reasonable to propose that instead of the current four thousand million, a population of half that would be appropriate for the long-term survival of a well-fed

Extremes of poverty and wealth are seen between countries and within countries. An example of the latter is shown here in the great disparity between the housing of the inhabitants of the shanty town in the foreground and the luxurious apartments of the rich on the skyline of Sao Paulo, Brazil.

world population. Such a prospect is a very long way off, even if every country agreed on its good sense. But, as the 1974 World Population Conference demonstrated all too clearly, global control of population is fraught with even more snares than is the problem of energy. Quite apart from the social and cultural barriers that continue to thwart national population plans, international politics could be as much a hinderance to global population control as a vehicle for its promulgation.

For instance at the 1974 conference in Bucharest, proposals by the rich nations for cutting down the rate of population growth were interpreted by a number of the poorer nations as merely another facet of imperialism – population imperialism. Who can blame them? Experience has taught these nations to be wary of the intentions of the affluent world. Unless the accumulated layers of suspicion that separate nations of different economic status and different political ideology can be cut through and stripped away, however, the path for global decision-making will remain blocked. And if that path does remain blocked, the threat to the continued survival of mankind becomes very real indeed.

From day to day, life on earth may seem in good enough order (provided one is a member of the affluent minority!) but in the perspective of biological time we have to face the fact that one day humanity *will* disappear. There is no escaping that fact. The question is, when?

At its most dramatic, the end of the planet earth could take place as a cosmological accident, sending all its passengers into oblivion. If such an event were to overtake us, that would be simply bad luck on a grand scale. More to the point is the question, can mankind thrive for a respectably long tenure without being destroyed by its own hand? The species to which we all belong, *Homo sapiens sapiens*, is perhaps fifty thousand years old, a mere infant in biological terms. It is quite possible that the path of evolution that brought us to our special position in the animal world, a position in which our inventiveness and culture allow us to manipulate our environment to an unprecedented degree, is soon to run into a dead end. Will the evolutionary step that handed over to an animal so much power and control over its environment turn out to have been the greatest biological blunder of all time? Can it be that the creation of the human species carried with it the seeds of ultimate destruction?

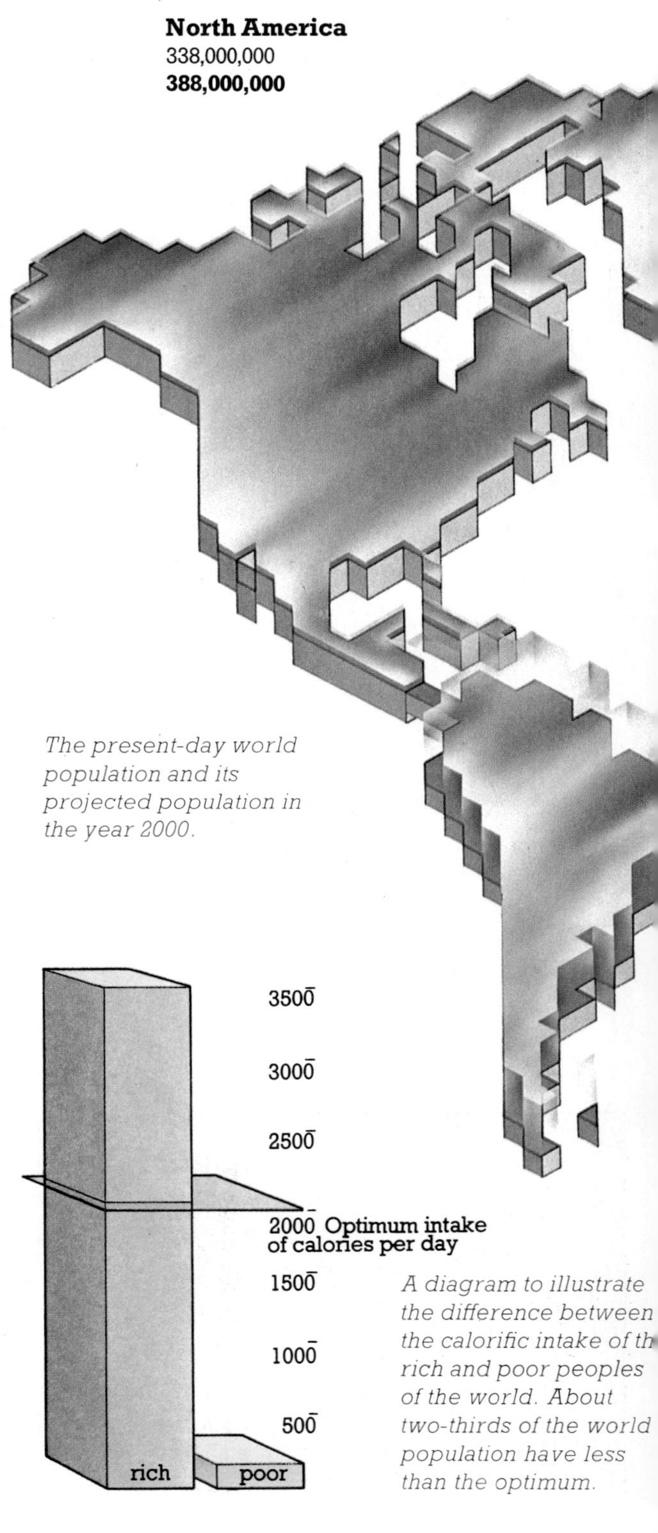

The present-day world population and its projected population in the year 2000.

A diagram to illustrate the difference between the calorific intake of th rich and poor peoples of the world. About two-thirds of the world population have less than the optimum.

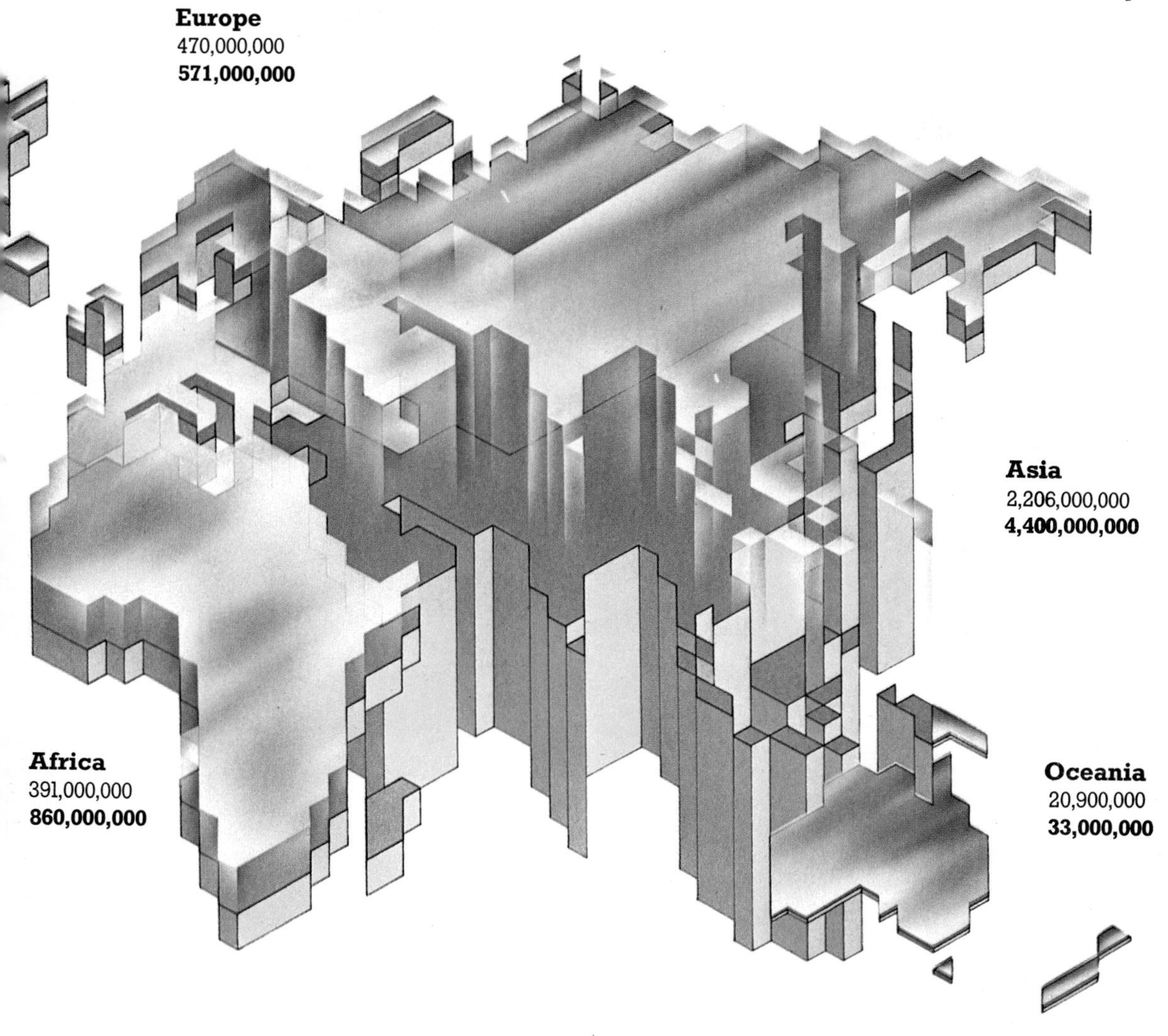

The world's material resources are, of course, finite, and one way of conserving them is to limit population growth. The birth rate appears to decrease as affluence increases, and there is an effort among the less well-endowed nations to limit population growth by means of birth-control education. 'Family Planning is Fine' (left) is a gouache (about 31 × 47 inches) by Sung Hou-cheng, a commune member of Hu county in Shensi, China. It depicts a doctor on a worksite with a portable exhibition explaining family planning.

The future of the human species depends crucially on two things: our relationships with one another, and our relationship to the world around us. The study of human origins can offer important emphasis in the way we view these two issues.

First, we are one species, one people. Every individual on this earth is a member of *Homo sapiens sapiens*, and the geographical variations we see among peoples are simply biological nuances on the basic theme. The human capacity for culture permits its elaboration in widely different and colorful ways. The often very deep differences between those cultures should not be seen as divisions between people. Instead, cultures should be interpreted for what they really are: the ultimate declaration of belonging to the human species.

It is a truism to say that politics are international. But this being so, it follows that any attempt to achieve long-term stability for humanity can come only through a global determination and will. It is not our intention to suggest how global politics might be run, with a world government or whatever other machinery might be appropriate. Rather, we wish to suggest that unless there is an acceptance of the oneness of the human race, a real spirit of brotherhood, then the political machinery, however sophisticated, will grind to a halt. The deep human drive for cooperation lends itself to achieve that aim. Just as the human propensity for group cooperation has in the past been harnessed to wage war between nations, it is now imperative that the same basic drive be channeled into a global effort to rescue humanity from itself.

An evolutionary perspective of our place in the history of the earth reminds us that *Homo sapiens sapiens* has occupied the planet for the tiniest fraction of that planet's four and a half thousand million years of existence. In many ways we are a biological accident, the product of countless propitious circumstances. As we peer back through the fossil record, through layer upon layer of long-extinct species, many of which thrived far longer than the human species is ever likely to do, we are reminded of our mortality as a species. There is no law that declares the human animal to be different, as seen in this broad biological perspective, from any other animal. There is no law that declares the human species to be immortal.

Unquestionably, mankind *is* special, and in many ways too. In past history, animals became extinct because for whatever reason the environment that had previously nurtured their birth turned hostile: perhaps the climate shifted, or competition from other species became too severe. Humans are the first animals capable of manipulating the global environment to a substantial degree. So, although through our evolution we have escaped many of the vicissitudes of the natural environment by becoming in some measure independent of it, we now have in our hands the engines of our own destruction. There is now a critical need for a deep awareness that, no matter how special we are as an animal, we are still part of the greater balance of nature. Unless we achieve such awareness the answer to the question of when the human species might disappear will be: 'sooner rather than later'.

Make no mistake: the human species *is* capable of gaining a keen awareness of its place in the global domain; and there is absolutely no reason why the world's population cannot operate in sympathy and harmony – and humility too – with the planet on which we so recently evolved. During that relatively brief span evolutionary pressures forged a brain capable of profound understanding of matters animate and inanimate: the fruits of intellectual and technological endeavour in this latter quarter of the twentieth century give us just an inkling of what the human mind can achieve. The potential is enormous, almost infinite. We can, if we so choose, do virtually anything: arid lands will become fertile; terrible diseases will be cured by genetic engineering; touring other planets will become routine; we may even come to understand how the human mind works!

No one with even the merest whiff of imagination would deny these predictions. What is at issue is whether nations can live peaceably with nations and with an understanding and deep respect for the natural world they inhabit so that one day these, and other, predictions may be fulfilled.

The answer, emphatically, is Yes. We certainly have the intellectual equipment with which to achieve it. And although the cultural spiral propelled successively by the Agricultural, Industrial, and Technological Revolutions is now spinning at a dizzying rate, we can clear our heads and measure short-term goals against the perspective of long-term existence. We are One People, and we can all strive for one aim: the peaceful and equitable survival of humanity.

To have arrived on this earth as the product of a biological accident, only to depart through human arrogance, would be the ultimate irony.

For Further Reading

Barash, David, **Sociobiology and Behaviour** (Oxford 1977)

Bender, Barbara, **Farming in Prehistory: from Hunter-Gatherer to Food Producer** (London 1975)

Bicchieri, M. G., **Hunters and Gatherers Today: A Socioeconomic Study of Eleven Such Cultures in the Twentieth Century** (New York 1972)

Dobzhansky, Theodosius, **The Biology of Ultimate Concern** (New York 1969; London 1971)

Robin Fox, **Biosocial Anthropology** (London and New York 1975; Association of Social Anthropologists, Series No. 1)

Geertz, Clifford, **The Interpretation of Cultures** (New York 1973; London 1975)

Harnard, Stevan, Horst Steldis, and Jane Lancaster (eds), 'Origins and Evolution of Language and Speech', **Annals of the New York Academy of Sciences,** Vol. 280 (New York 1976)

Jerison, Harry, **Evolution of the Brain and Intelligence** (New York 1973; London 1974)

Jolly, Alison, **The Evolution of Primate Behavior** (New York and London 1972)

Kessler, Evelyn, **Women: An Anthropological View** (New York 1976)

Lee, Richard B. and Irven De Vore (eds), **Man the Hunter** (Chicago 1968)

Lee, Richard B. and Irven De Vore (eds), **Kalahari Hunter-Gatherers: Studies of the Kung Sang and their Neighbors** (Cambridge, Mass., 1976)

Montagu, M. F. Ashley, **The Nature of Human Aggression** (New York 1976)

Rumbaugh, Duane, **Language Learning by a Chimpanzee** (New York and London 1977)

Wilson, Edward O., **Sociobiology: The New Synthesis** (Cambridge, Mass., 1975)

Magazines

Accounts of new fossil finds can be read in **National Geographic Magazine** (Washington, D.C., monthly) and **New Scientist** (London, weekly).

Acknowledgments

The producers would like to thank all those picture agencies, institutions and persons listed below for providing the illustrations used in the book. Many provided so much material that choice was difficult. In particular we should like to express our gratitude to the following people who willingly passed on a great deal of information, at no little cost in time, without which many of the drawings could not have been compiled: Richard Gray, Alison Cooke, Dr Peter Andrewes, Dr Christopher Stringer, Dr Todd Olsen, Dr P Napier.

Bold type denotes drawings, diagrams and maps.

6–7 Photo: The Hale Observatories (Mt Wilson and Palomar). Marshall Cavendish Picture Library
8 *Top left* © National Geographic Society (photo: Leakey Family Collection)
Left © National Geographic Society (photo: Gilbert M. Grosvenor)
8–9 *Top* ZEFA, Dusseldorf (photo: P. Fera)
9 *Top* © National Geographic Society (photo: Hugo van Lawick)
Middle © National Geographic Society (photo: Mary Leakey)
Bottom Photo: Roger Lewin
10 *Top* Natural History Picture Agency, Westerham (photo: Ivan Polunin)
Bottom Bruce Coleman Ltd, London (photo: Lee Lyon)
11 *Far left* Photo: Peter Andrewes
Left AAA (photo: Germain Guillemot)
Bottom left Anthro-Photo, Cambridge, Mass. (photo: Irven De Vore)
Right Natural Science Photos, Watford (photo: G. Newlands)
Bottom right © National Geographic Society (photo: David Brill)
12–13 **Alison Cooke/Len Whiteman: Gordon Cramp Studios**
14–15 **Alison Cooke/Len Whiteman: Gordon Cramp Studios**
16–17 **Map by Gordon Cramp Studios**
18–19 Ann Ronan Picture Library, Loughton, Essex
20 The Bettmann Archive, Inc., New York
21 Ann Ronan Picture Library, Loughton, Essex
22–3 Ann Ronan Picture Library, Loughton, Essex
23 Radio Times Hulton Picture Library, London
24 *Top* The Mansell Collection, London
Bottom The Bettmann Archive, Inc., New York
25 *Top* The Mansell Collection, London
Bottom Mary Evans Picture Library, London
26 *Top* **Map by Gordon Cramp Studios**
Bottom left Photo: Derek Witty
Bottom right Photo: John Freeman
27 *Both pictures* Reproduced by courtesy of the Royal College of Surgeons of England
29 Radio Times Hulton Picture Library, London
30 The Mansell Collection, London
31 Mary Evans Picture Library, London
32 **Map by Gordon Cramp Studios**
33 *Left* Ardea Photographics, London
Right Courtesy of the Trustees of the British Museum (Natural History), London
34–5 **Ronald Bowen**
36–7 John Hillelson Agency, London (photo: Dr Georg Gerster)
38 Bruce Coleman Ltd, Uxbridge
39 *Top* **Ronald Bowen**
Bottom Bruce Coleman Ltd,

Uxbridge (photo: Helmut Albrecht)
40 Anthro-Photo, Cambridge, Mass. (photo: T. W. Ransom)
41 Anthro-Photo, Cambridge, Mass. (photo: James Moore)
42–3 **Ronald Bowen**
44 **Nigel Osborne**
46 *Both pictures* Marshall Cavendish Picture Library
47 Bruce Coleman Ltd, London (photo: G. D. Plage)
50 *Left* Natural History Picture Agency (photo: Ivan Polunin)
Right Bruce Coleman Ltd, London (photo: S. C. Bisserot)
Bottom right Bruce Coleman Ltd, (photo: Norman Myers)
50–1 **Map by Gordon Cramp Studios**
51 *Top left* Bruce Coleman Ltd, London (photo: Jane Burton)
Left Bruce Coleman Ltd, London (photo: Norman Tomalin)
Bottom left Bruce Coleman Ltd, London (photo: Lee Lyon)
Right Pitch, Paris (photo: J. M. Cresto)
52 Photo: Peter Andrewes
53 *Top* **Map by Gordon Cramp Studios**
Bottom **Richard Gray**
54 *Top* Photo: E. S. Ross, San Francisco
Bottom AAA Photo, Paris (photo: Jean-Claude Chabin)
55 *Left* Bruce Coleman Ltd, London (photo: M. P. Price)
55 *Right* Anthro-Photo, Cambridge, Mass. (photo: Russell A. Mittermeier)
56 Photos: Peter Andrewes
57 **Len Whiteman/Gordon Cramp Studios**
Bottom **Ronald Bowen**
58–9 Bruce Coleman Ltd, Uxbridge (photo: Dian Fossey)
60–1 John Topham Picture Library, Edenbridge (photo: Simon Trevor)
61 Anthro-Photo, Cambridge, Mass. (photo: Wrangham)
62 Bruce Coleman Ltd, Uxbridge (photo: Norman Tomalin)
63 Bruce Coleman Ltd, Uxbridge (photo: Mike Price)
65 John Topham Picture Library, Edenbridge (photo: Lee Lyon)
66 **Map by Gordon Cramp Studios**
67 *Left and center* Photos: Peter Andrewes
Right **Ronald Bowen**
68 John Topham Picture Library, Edenbridge (photo: Leonard Lee Rue III)
69 *All pictures* Photos: Timothy Ransom
70 Colorific! London (photo: Terence Spencer)
71 Anthro-Photo, Cambridge, Mass. (photo: Wrangham)
72–3 *Top* **Marian Appelton**
73 *Bottom* **Gordon Cramp Studios**
74 Photo: Peter Andrewes
75 Anthro-Photo, Cambridge, Mass. (photo: Nancy Nicolson)
76–7 **Marian Appelton**
78–9 **Maps by Richard Gray/Gordon Cramp Studios**
82 *All pictures* Photos: Diane Gifford
83 Photo: Diane Gifford
84–5 **Len Whiteman/Gordon Cramp Studios**
86 *All pictures* Bruce Coleman Ltd, Uxbridge (photos: R. I. M. Campbell)
87 © National Geographic Society (photo: R. I. M. Campbell)
89 *Both pictures* Photos: Diane Gifford
90 *Top and center* © National Geographic Society (photos: David Brill)
Bottom Cleveland Museum of Natural History
91 © National Geographic Society (photo: David Brill)
92–3 Photo: Peter Andrewes
93 *Top* **Map by Gordon Cramp Studios**
Bottom (both pictures) Photos: E. Delson
94 Shostal Associates, New York
95 **Ronald Bowen**
97 *Top* © National Geographic Society (photo: Melville Bell Grosvenor)
Bottom © National Geographic Society (photo: Robert M. Campbell)
98–9 Anthro-Photo, Cambridge, Mass. (photo: Canon)
99 *Top* **Map by Gordon Cramp Studios**
Bottom **Richard Gray**
100 *Left* **Ann Winterbotham**
100–1 Picturepoint Ltd, London
101 **Ann Winterbotham**
102 *Both pictures* Photos: E. S. Ross, San Francisco
104 John Hillelson Agency, London (photo: Dr Georg Gerster)
105 John Hillelson Agency, London (photo: Dr Georg Gerster)
106 *Left* **Ronald Bowen**
Right John Topham Picture Library, Edenbridge (photo: Des Bartlet)
107 *Left* © National Geographic Society
Right **Ronald Bowen**
109 *Top* © National Geographic Society
Bottom Photo: Diane Gifford
110 *Both pictures* Bruce Coleman Ltd, Uxbridge (photo: R. I. M. Campbell)
110–1 **Ronald Bowen**
112 *Top* Anthro-Photo, Cambridge, Mass. (Photo: R. Lee)
Bottom Photo: E. S. Ross, San Francisco
114–5 **Ronald Bowen**
118–9 Picturepoint Ltd, London
120–1 **Map by Gordon Cramp Studios**
122 **Ronald Bowen**
123 **Ronald Bowen**
126–7 **Ronald Bowen**
128 *All pictures* Photos: Ralph S. Solecki
129 *Left (both pictures)* Courtesy Henry De Lumley
129 *Right* **Marian Appelton**
130 **Ronald Bowen**
131 Photo: C. Stringer
132 Courtesy of the American Museum of Natural History, New York
133 Courtesy of the American Museum of Natural History, New York
134–5 **Ronald Bowen**
136 *Top* Photo: C. Stringer
Bottom Photo: E. Delson
138 *Top left* Daily Telegraph Colour Library (photo: L. L. T. Rhodes)
Top center Bruce Coleman Ltd, Uxbridge (photo: Norman Myers
Top right Picturepoint Ltd, London
Bottom Colorific! London (photo: John Moss)
139 *Top* Colorific! London (photo: John Moss)
Left **Nigel Osborne**
Bottom Photo: E. S. Ross, San Francisco
140 *Left* Michael Holford Library
Right Agence Hoa-Qui, Paris

141 *Both pictures* **Richard Gray**
142 **Ronald Bowen**
143 **Gordon Cramp Studios**
146–7 Photo: E. S. Ross, San Francisco
149 **Len Whiteman/Gordon Cramp Studios**
150 *Top* Agence Hoa-Qui, Paris
Bottom Anthro-Photo, Cambridge, Mass. (photo: Irven De Vore)
151 Anthro-Photo, Cambridge, Mass. (photo: Irven De Vore)
152 Ardea Photographics, London (photo: Gert Bekrens)
153 *Both pictures* John Topham Picture Library, Edenbridge (photo: Norman Myers)
155 *Top* Anthro-Photo, Cambridge, Mass. (photo: Irven De Vore)
Bottom Anthro-Photo, Cambridge, Mass. (photo: Jiro Tanaka
156 *Top* Pitch, Paris (photo: P. Montoya)
Bottom Bruce Coleman Ltd, Uxbridge (photo: S. Pearson)
158–9 © National Geographic Society (photo: Robert Campbell)
160 **Map by Gordon Cramp Studios**
161 Picturepoint Ltd, London
162–3 Anthro-Photo, Cambridge, Mass. (photo: Washburn)
164 *Both pictures* Anthro-Photo, Cambridge, Mass. (photos: Richard Lee)
165 Anthro-Photo, Cambridge, Mass. (photo: Irven De Vore)
166 Photo: E. S. Ross, San Francisco
167 *Top and lower right* Photos: E. S. Ross, San Francisco
Left Anthro-Photo, Cambridge, Mass. (photo: Richard Katz)
168 Anthro-Photo, Cambridge, Mass. (photo: Richard Katz)
169 Anthro-Photo, Cambridge, Mass. (photo: Richard Lee)
170 Anthro-Photo, Cambridge, Mass. (photo: Irven De Vore)
171 Anthro-Photo, Cambridge, Mass. (photo: Irven De Vore)
173 Shostal Associates, New York
174 Anthro-Photo, Cambridge, Mass. (photo: Irven De Vore)
175 Anthro-Photo, Cambridge, Mass. (photo: Irven De Vore)
176 Robert Harding Associates, London
177 **Map by Richard Gray/Gordon Cramp Studios**
178–9 Bruce Coleman Ltd, Uxbridge (photo: Manfred Kage)
180 Michael Holford Library, London
181 *Both pictures* Fox Photos Ltd, London
182 John Hillelson Agency, London (photo: Howard Sochurek)
183 Daily Telegraph Colour Library, London
184 *All pictures* Photos: Michael Lyster
186 **Alison Cooke/Nigel Osborne**
187 **Alison Cooke/Nigel Osborne**
190 Ardea Photographics, London (photo: Dr P. Morris)
191 **Alison Cooke/Nigel Osborne**
193 **Alison Cooke/Nigel Osborne**
194–5 **Ann Winterbotham**
195 Bruce Coleman Ltd, Uxbridge (photo: C. James Webb)
196 John Hillelson Agency, London
198–9 **Richard Gray/Gordon Cramp Studios**
199 *Left* © National Geographic Society (photo: Mary Leakey)
Right © National Geographic Society (photo: Hugo van Lawicke)
201 *Both pictures* Colorific! London (photo: Nina Leen)
202 Bruce Coleman Ltd, Uxbridge (photo: R. R. Murton)
203 *Top* **Nigel Osborne**
Bottom (both pictures) Photos: Ralph Holloway
206–7 Sunday Times, London (photo: Donald McCullin)
209 *Top* Bruce Coleman Ltd, Uxbridge (photo: Jane Burton)
Bottom Bruce Coleman Ltd, Uxbridge (photo: D. and K. Urrey)
210 *Top* John Topham Picture Library, Edenbridge (photo: Bob Campbell, Armand Denis Productions)
Bottom John Topham Picture Library, Edenbridge (photo: D. Quignard)
210–1 Susan Griggs Agency, London (photo: John Garrett)
211 John Hillelson Agency, London (photo: Eve Arnold)
214 Bruce Coleman Ltd, Uxbridge (photo: Leonard Lee Rue III)
214–5 Photri, Alexandria, Va. (photo: L. Novak)
216–7 Keystone Press Agency, London
218–9 Daily Telegraph Colour Library, London (photo: Leonard Rhodes)
222 Sunday Times, London (photo: Donald McCullin)
226–7 Bruce Coleman Ltd, Uxbridge (photo: Francisco Futil)
228 Camera Press, London (photo: Marian Kaplan)
230–1 Explorer, Paris (photo: Jean Valentin)
232 Shostal Associates, New York
234 John Hillelson Agency, London (photo: Ian Berry)
235 *Top* Colorific! London (photo: Tony Carr)
Bottom Susan Griggs Agency, London (photo: Englebert)
236 Anthro-Photo, Cambridge, Mass. (photo: R. Lee)
238–9 Space Frontiers Ltd, Havant (photo: N.A.S.A.)
240–1 Keystone Press Agency Ltd, London
242 John Hillelson Agency (photo: Zetas/Duttilleux)
243 *Top* Royal Society/Royal Geographical Society (photo: Hugh I. Jones)
Bottom E. S. Ross, San Francisco
244 John Hillelson Agency, London (photo: Ernest Cole)
245 *Top* Bruce Coleman Ltd, Uxbridge (photo: Masod Qureshi)
Bottom Robert Harding Associates, London
246 *Left* Bruce Coleman Ltd, Uxbridge (photo: Brian J. Coates)
Bottom Bruce Coleman Ltd, Uxbridge (photo: Nicolas Devore)
246–7 Picturepoint Ltd, London
247 *Top* Bruce Coleman Ltd, Uxbridge (photo: Brian J. Coates)
Left Bruce Coleman Ltd, Uxbridge (photo: Christian Zuber)
Bottom Bruce Coleman Ltd, Uxbridge (photo: Fritz Prenzel)
250 *Top* Explorer, Paris (photo: Jacques Trotignan)
Left Bruce Coleman Ltd, Uxbridge (photo: Chris Bonington)
Right Picturepoint Ltd, London
251 *Top* Explorer, Paris (photo: H. Veiller)
Bottom Colorific! London (photo: David Moore)
252–3 Colorific! London (photo: John Moss)
254 **Nigel Osborne**
254–5 **Richard Gray/Gordon Cramp Studios**
255 Arts Council of Great Britain

Index

References in *italic* type indicate illustrations or material displayed in 'boxes': the page numbers are those of the captions, not of the illustrations themselves. A figure 2 in brackets immediately after a page number indicates that there are two separate references to the subject on that page. Footnotes are indicated by '*n.*' following the page number.